THE ULTIMATE GUIDE TO



# BUTCHERING, SMOKING, CURING, SAUSAGE, *and* JERKY MAKING

# THE ULTIMATE GUIDE TO BUTCHERING, SMOKING, CURING, SAUSAGE, and JERKY MAKING

Philip Hasheider

**Quarto.com**



First Published in 2019 by
The Harvard Common Press,
an imprint of The Quarto Group,
100 Cummings Center, Suite 265-D,
Beverly, MA 01915, USA.
T (978) 282-9590
F (978) 283-2742



The Harvard Common Press titles are also available at discount for retail, wholesale, promotional, and bulk purchase. For details, contact the Special Sales Manager by email at specialsales@quarto.com or by mail at The Quarto Group, Attn: Special Sales Manager, 100 Cummings Center, Suite 265-D, Beverly, MA 01915, USA.

ISBN: 978-1-55832-987-4

Digital edition published in 2019
eISBN: 978-1-55832-988-1

Library of Congress Cataloging-in-Publication Data available.



Design and layout: Burge Agency
Photography: Photos courtesy of Marcus Hasheider, except for those from Creative Publishing International (pages 30 [right top and bottom] and 117 to 127) and Shutterstock (pages 6–7, 9, 10, 12 [top left and bottom left and right], 24, 38, 39, 62, 63, 78, 79, 96, 97, 114, 115, 128, 129, 140, 141, 146, 148, 155, 156, 157 [left], 158 [top], 160 [right], 161, 162 [left and middle], 165, 180 [left], 182, 183)

THIS BOOK IS DEDICATED TO
THE NEXT GENERATION OF OUR
FAMILY: MARCUS AND PAIGE;
AND JULIA AND VICTOR.

# CONTENTS

# INTRODUCTION

**INTEREST IN HOME BUTCHERING HAS DRAMATICALLY INCREASED IN RECENT YEARS. THIS INTEREST IN BUTCHERING HAS MANY VARIABLES AND MAY BE AS DIVERSE AS THOSE WHO WISH TO DO IT.**

In *The Ultimate Guide to Butchering, Smoking, Curing, Sausage, and Jerky Making*, you will find specific and detailed instruction that is most applicable to the task at hand that is accurate and that will help assure an end product that is safe to be consumed by you and those you invite to your table.

This book will help you get started, do it safely, and do it accurately. Safety during the butchering process cannot be overstated. This book will offer you many safety tips and alert you to where potential problems may surface. It will also help you recognize those areas so that you can take the steps necessary to ensure these problems do not arise.

You'll also be guided through the entire processes of transforming a live animal into portions that appear on your plate. Chapters are dedicated to specific species. Whether it is beef, pork, sheep, chicken, or deer, this book will take you through the step-by-step protocol in transforming the carcass into edible portions.

Along the way, you will learn all the necessary sanitation steps required to maintain a safe environment for the meat. You will learn how to safely use, handle, and protect your knives, as well as protect yourself and others who may help. You will learn all the steps necessary for preserving your meat products by transforming them into sausages, jerky, or other dry meats.

You may also learn something about yourself through this process: that it is an age-old dynamic between animals and the humans who raise and care for them. You may develop a greater appreciation for all the work it takes to bring an animal to your table. Learning to utilize *every* part of a carcass means that nothing goes to waste. If we can develop an atmosphere where a consumer wants all of the animal, it can create the ultimate form of gratitude to the animal being used: Nothing is wasted.

Human civilization has consumed animal meat products as part of its diet for thousands of years. Today's consumers can choose from a variety of animal species, such as beef cattle, to include as part of their daily diet.

## HOW TO USE THIS BOOK

The purpose of this book is to provide the reader with accurate and useful information on all aspects of butchering, from slaughtering to processing to the preservation of meat. In these pages, you will find all the fundamental processes involved with safely handling beef, pigs, sheep, and deer. Detailed step-by-step instructions—from securing an animal to deconstructing the carcass—are found within each species chapter that will allow you to safely and humanely transform that particular carcass into family meals. The identification and names of different meat cuts are discussed in each section to help you understand the location and purpose of each cut.

There is also much information on smoking and preserving meat. This makes sense, as with whole-animal butchering, you are often faced with many more pounds (kilograms) of meat than you can reasonably eat fresh—or sometimes even freeze. To that end, this book contains chapters on smoking, curing, sausage, and jerky making to help you safely preserve your meat for future use.

CHAPTER 1

# MUSCLES ARE MEAT

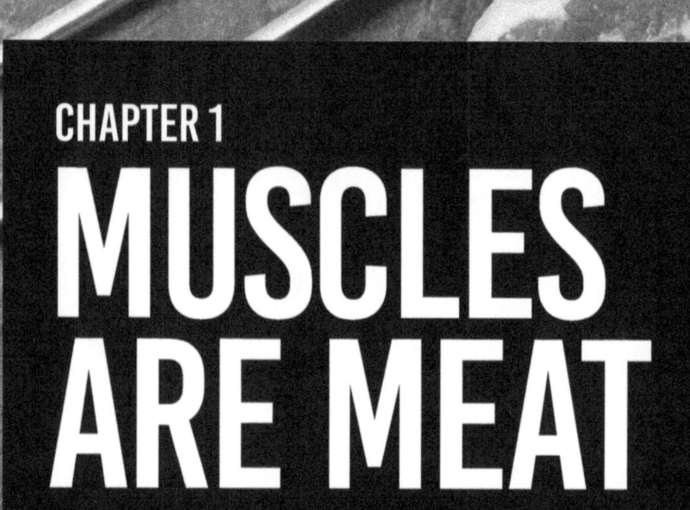

**IN THE PAST, THE INTERNAL ORGANS OF ANIMALS WERE A HIGHLY PRIZED PART OF THE CARCASS. ALTHOUGH MANY ARE STILL USED TODAY IN A VARIETY OF FOOD, PHARMACEUTICAL, AND HEALTH CARE APPLICATIONS, THEIR VALUE HAS BECOME SECONDARY TO THAT OF THE MUSCLES. THE DECREASE IN SELECTION CHOICES HAS RESULTED FROM THREE MAJOR CHANGES IN OUR SOCIETY, NAMELY CULTURAL CONDITIONING, INDUSTRIAL URBANIZATION, AND THE DIFFICULTY IN DOMESTICATING SOME WILD SPECIES AND/OR THEIR SUPPLY.**

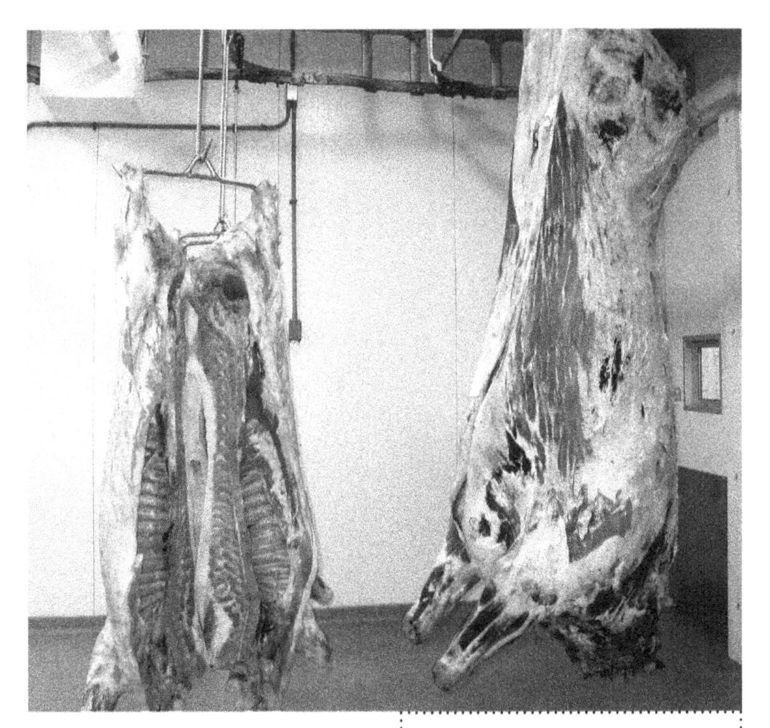

Beef carcasses can reach and exceed seven feet (2 m) in length after hanging and pork carcasses five feet (1.5 m). If you hang them for aging, they must not touch the floor or ground to avoid contamination.

Cultural conditioning resulted from a lack of variation in available food choices, particularly from wild game. What was familiar and more readily available became the norm of what was eaten, moving away from traditional foods. The movement of rural residents to urban centers during the industrialization period of this country resulted in having to purchase meat from a local market rather than procuring it from your own efforts in raising animals.

Some wild species, such as bison, elk, and antelope, were difficult, if not impossible, to domesticate or procure, resulting in fewer being used for meat purposes.This evolution of eating preferences or choices has not gone unnoticed. The United States Department of Agriculture (USDA) reported in 2011 that per capita total meat consumption in the United States rose from 104.9 pounds (47.6 kg) in 1970 to 110.4 pounds (50 kg) in 2009.

Geese, ducks, and poultry are among the fowl that can provide a distinctive meal for special occasions. Small in size, they require less feeding than large animals and reach target butchering weights in short time periods.

It does not require owning a large acreage to raise meat animals. Small pastures of one to two acres (4,047 to 8,094 m$^2$) would be sufficient to raise several meat goats or sheep for home use.

Ring-necked pheasants are one type of wild game bird that can be used for meat dishes. Game birds are regulated compared to domestically raised fowl, which are not. You should check state rules and regulations before hunting.

Heritage beef breeds, such as the Scottish Highland, can be grown for home use. Different species and breeds offer different carcass characteristics and also have different habitat adaptabilities. You should study which ones may suit your situation best.

# TODAY'S MEAT ANIMALS

The animals we use in our diets are either herbivores, such as cattle and sheep, or more or less omnivores, such as pigs and chickens. The texture of the muscles and the fats found in the bodies of different species are largely reflective of their diet. Genetic factors are also involved. Grass-fed domestic animals are the only group that has a fat profile that is similar to wild herbivores. Supplements of grains or compounded feeds that are fed to domestic animals will change this fat profile in their bodies to reflect the fats found in those plants. For example, pigs are often fed high grain and soybean diets to produce rapid growth. This produces fat profiles that are typically high in oleic acids and lower in palmitic acids. Because pigs are omnivores, their muscle fat reflects the kinds of fat they are fed. Cattle, however, are natural grass eaters if given the opportunity, and their body fat profile will reflect the influence of plant nutrients.

Consumers are becoming more aware of differences in fat composition, production, and their effects on human health. Grain-fed cattle, because of their high corn and soybean diets, have a higher omega-6 and lower omega-3 fat profile than their counterparts that are grass-fed. The latter typically have 7 percent of omega-3; wild animals can reach a level of about 4 percent. However, saturated fats have a purpose. They aid in meats retaining their quality because they are less subject to oxidation. Highly unsaturated fats are soft and oily and may lower the quality of pork and poultry carcasses because they are more readily oxidized and may reduce shelf life. Oxidation produces off-flavors.

A result of these differences can be seen in the appearance of the cuts. Those from a high fat diet, such as corn and soybeans, will typically have more fat in the body of the carcass as well as in the muscling, commonly referred to as marbling. The white streaks interlaced in different cuts will change the texture of the muscle.

Grass-fed beef has a higher antioxidant capacity than feedlot beef. This means grass-fed meat cuts retain their red color longer without artificial manipulation.

What does this mean for you? If you are purchasing an animal to butcher yourself, it may provide further information to make a more informed choice and alert you to what you may encounter in cutting up a carcass.

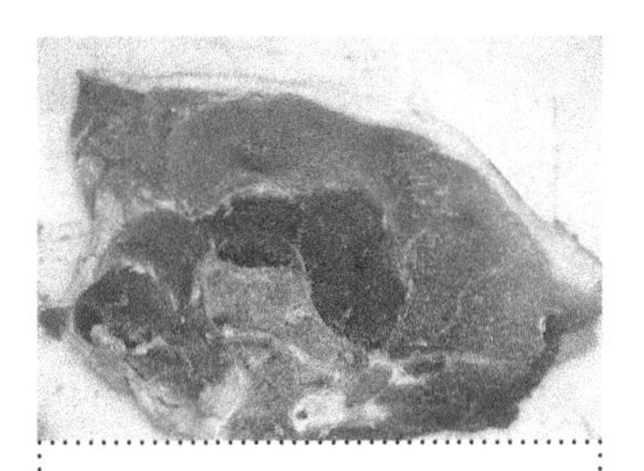

Marbling refers to the white specks of fat within a muscle and seam fat is the streaks that surround the muscle. Fat amounts may vary among animals of the same species due to diet and genetics. Large amounts of fat should be trimmed off the carcass but can be useful for rendering into lard or mixed with wild game to add texture and flavor to sausages.

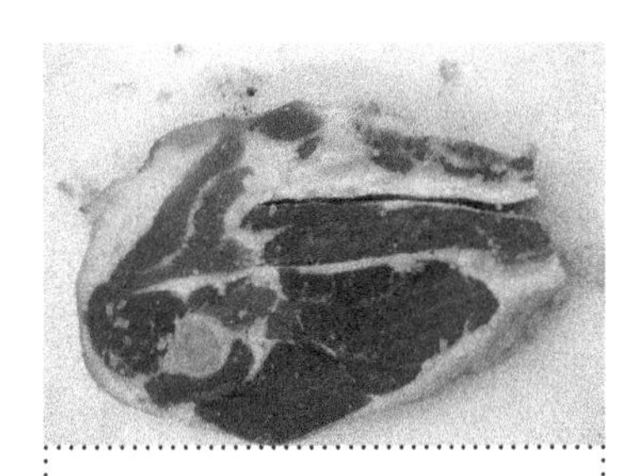

Well-grown lambs have a good layer of fat covering many of their prime cuts and in areas between the muscles. Most of the portions of fat around the exterior part of the cuts should be trimmed.

# CONVERTING MUSCLE TO MEAT

Changes happen within the muscles of an animal after it is slaughtered or harvested. Some life processes stop almost immediately. Other life processes gradually cease over a slightly longer, but finite, time period.

Although other dramatic changes occur initially, such as the loss of brain activity, heart action, lung function, blood transport, digestive action, and mobility, the changes occurring within the muscles can ultimately affect the eating quality of the meat.

When an animal is killed or dies in other circumstances, the muscle pH gradually drops. This results because the animal's glycogen reserves within the muscle are depleted and they are converted to lactic acid. Oxygen is no longer available to the muscle cells after the animal is bled, causing a lactic acid buildup and a subsequent drop in pH.

With the loss of certain muscle reserves such as creatine phosphates, which help in muscle movements, the muscle filaments can no longer slide over one another and the muscle becomes still and rigid, resulting in a condition known as rigor mortis.

Several factors influence the amount of time for the muscle to reach its final pH level. These include the species, cooling rate, and the extent of the animal's struggle at the time of death.

Beef and lamb muscles take longer to reach their final pH than those of a pig. Cooling affects the time because metabolism is slowed when the carcass is subjected to lower temperatures. Finally, the animals' activity level immediately prior to the killing will affect the pH; less activity will prolong the period of pH decline.

During the period after the slaughter or harvest, changes also occur in the muscle proteins as they begin to break down. This generally occurs during the cool storage period and is referred to as "aging" and results in increased meat tenderness. This process of protein fracturing will continue for one to two weeks, after which there is little appreciable increase in tenderness.

Oxidation, coupled with surface drying (dehydration of the proteins), causes a change in meat color because of the interaction between oxygen and myoglobin. Myoglobin is a meat color pigment that gives the meat its red color. When it binds with oxygen molecules, the color deepens. The longer muscles are exposed to air during the aging process, the more likely a change in surface color. You can cover exposed muscle with plastic wrapping to minimize its effects, or you can slice off affected portions.

## MEAT QUALITY CHALLENGES

For generations, people butchering animals have known the importance of harvesting healthy animals that have not been made excitable or stressed immediately prior to slaughter. If an animal undergoes vigorous stress or exercise before harvest, the glycogen content within the muscles may drop dramatically. This can result in a higher pH remaining in the muscles, causing the meat to become dark, firm, and dry—effectively reducing the tenderness and quality.

High pH meat typically is dark in color, believed to be the result of a greater water-holding capacity, which causes muscle fibers to swell. The meat from such animals generally has a reduced shelf life because a higher pH is more likely to accommodate bacterial growth.

A second quality problem can result because of the rate of pH decline. If the muscle pH drops too rapidly after killing, due to the muscle temperature being too high, it can become pale and soft. This results in a soft, mushy texture, a pale color, and the muscle lacking the ability to hold moisture. This condition typically results from high-stress situations but also, in the case of some pigs, can be a hereditary stress condition resulting from porcine stress syndrome (PSS). The muscle temperature can be affected by animal excitement initiating a "fright or flight" response, further stimulating nervousness and sweating.

While there are meat-quality issues associated with harvest and post-harvest handling of carcasses, quality and food safety issues can affect animals at pre-harvest. These include animal health, pregnancy, injury and bruising, and genetic influences. Anyone

Studying the pig's anatomy before slaughter and butchering will help you understand the internal organ placement within the body cavity. This will be useful, particularly when butchering a pig for the first time.

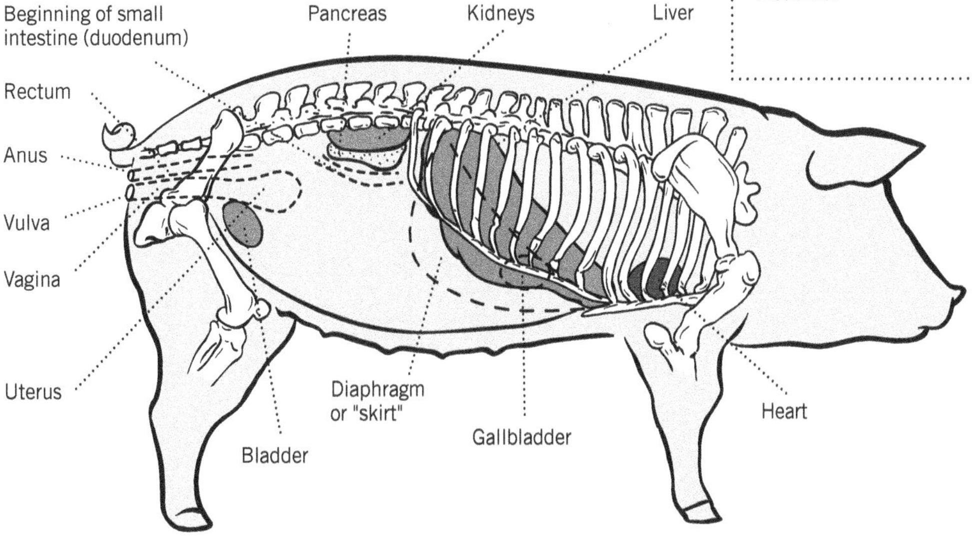

A beef animal has a much larger body structure than a pig, lamb, or most wild game animals. This drawing identifies areas for helping handle the carcass after it has been put down and the location of the internal organs in relation to the skeletal structure.

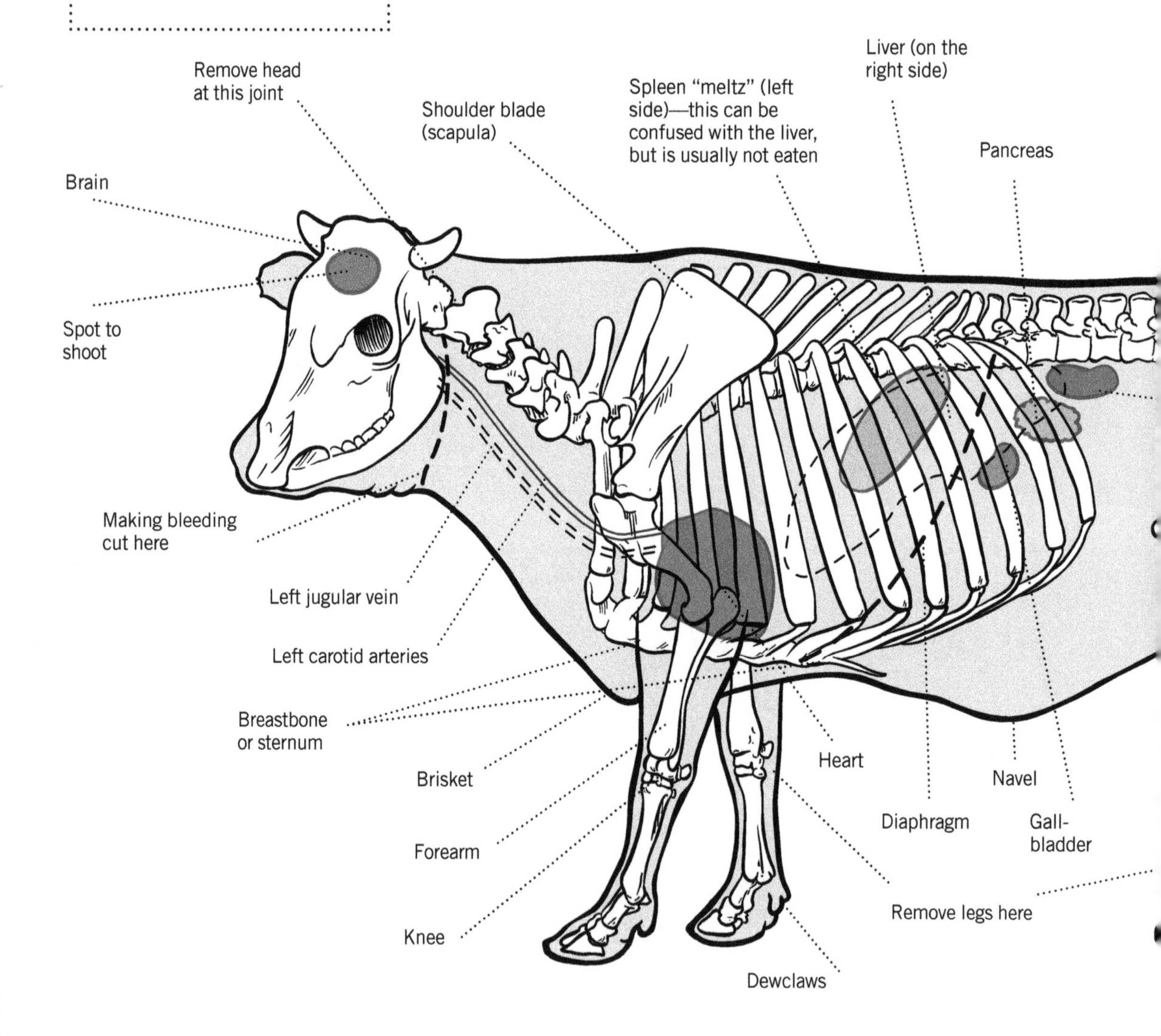

performing a harvest for home use or for sale should be aware of these factors. Recognizing unhealthy or unthrifty animals can help avoid subsequent meat quality problems. Common ailments can affect the quality, value, and wholesomeness of a carcass. If you are unsure about the health of an animal to be harvested for your own use, ask someone who is knowledgeable or consult a veterinarian.

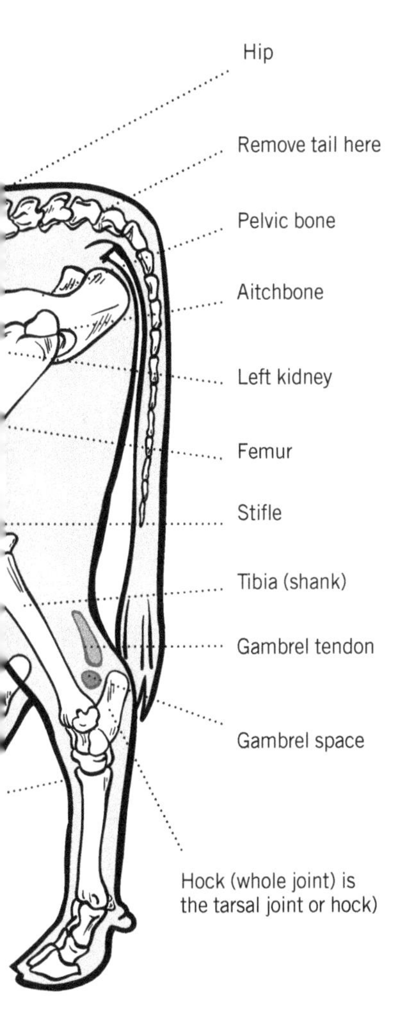


Animals that have received therapeutic protocols, such as injections of antibiotics or hormones, are required to be withheld from the human food chain for a specific period of time, depending on what is administered. It is extremely important to understand and abide by any withdrawal times associated with antibiotic use, and it is illegal to send any animal to market prior to the expiration of the specified withdrawal period.

Avoid harvesting a pregnant animal, particularly one in the last trimester. This period is characterized by an increase in the hormonal level, including oxytocin and estrogen. These may affect the entire animal and not simply specific parts. Also, the body temperature of a pregnant female may elevate slightly during this period prior to or after parturition. However, once her body temperature returns to normal and stabilizes at about 101°F (38.5°C), in the case of a beef cow, you can proceed with a harvest. Similar conditions are found in other species, so you will need to apply the same logic to whatever female animal you will be using for meat.

Animals that are healthy but have been injured because of an accident, such as a broken leg or pelvis, may be used for harvest but preferably as soon as possible. An injured animal easily becomes nervous or agitated, especially if it cannot move about as it was generally accustomed to doing. Pain from an injury may also influence its behavior. Although pain may influence the animal's movements or lack of them, it does not influence the quality of the meat by itself; it becomes a secondary contributor to meat quality. If an animal dies due to some nonviral condition such as bloat, where the gas formed in the digestive system has no way to escape and slowly builds pressure against the lungs to the point where they cease to function, or from a heart puncture, they can be used for meat, providing a knowledgeable person is present to immediately cut the throat and bleed the animal. Under no circumstances should a dead or diseased animal be used for human food.

Bruising is another condition that can affect meat quality. Bruises result from the hemorrhaging of blood vessels under the hide of the animal. You should not use any bruised meat for food. Most bruises occur during the loading and unloading of animals to transport them to the point of harvest. In cattle, the major sites for potential bruising are in the loin and sirloin area, ribs, and shoulders. In swine, the highest percentage of bruising occurs in the ham, shoulders, and loin. Sheep experience the highest percentage of bruising in the legs and loin due to grabbing them by the wool or catching them by the hind leg.

## FACTORS AFFECTING MEAT SAFETY

Other factors that can affect meat quality include toxins, bacteria, viruses, and temperature, time, and moisture. For many years, it was thought that the muscle of an animal was sterile if it had not been injured, cut into, or bruised. In recent years, however, researchers have found viable bacteria within muscle tissue. This means that when you harvest an animal, whether domestic or wild, extreme care must be taken to prevent the introduction of foreign bodies into the carcass through your actions. This care begins with the knives you use to sever the jugular vein at the beginning of the slaughtering process and continues until the cuts have been packaged, sealed, and stored—or the meat is immediately cooked for use. Sanitation is extremely important and is discussed in greater detail elsewhere in this book.

Preventing and retarding the development of harmful microorganisms should be your primary objective in harvesting your animal or in-home processing any meat products. Consuming microorganisms that have grown and propagated in meat can cause illness or even death. This concern should not be taken lightly. When health problems arise related to eating meat, it is generally a result of intoxication or infection. Intoxication occurs when the microbe produces a toxin that is subsequently eaten by a human and sickness results. Infection occurs when an organism is eaten by a human, then grows and disrupts the normal functions of the body, such as salmonella and listeria.

There are several types of toxins, including exotoxins and endotoxins. Exotoxins are located outside the bacterial cell and are composed of proteins that can be destroyed by heat through cooking. Exotoxins are among the most poisonous substances known to humans. These include *Clostridium botulinum*, which causes tetanus and botulism poisoning.

Endotoxins attach to the outer membranes of cells but are not released unless the cell is disrupted. These are complex fat and carbohydrate molecules, such as *Staphylococcus aureus*, that are not destroyed by heat.

Bacteria are the most common and important microorganisms that can grow on meat. Not all bacteria are bad, however, as the human body may carry as many as 150 different kinds of bacteria on it.

Molds and yeasts are fungi that can affect meat quality, although their effect is far less significant or life threatening than toxins or bacteria. Molds typically cause spoilage in grains, cereals, flour, and nuts that have low moisture content and in fruits that have a low pH. Yeasts are generally involved where a food product contains high amounts of sugar. Yeasts that affect meat are generally not a problem because of the low sugar or carbohydrate content of muscle.

Viruses, while having the potential to cause food diseases, generally only affect raw or uncooked shellfish. Viruses are inert and unable to multiply outside a host cell.

There are a few parasites that you should be aware of that may cause problems in meat. A parasite infection will occur in the live animal before it occurs in a human. There are three parasites that are of major concern to humans: *Trichinella spiralis*, *Toxoplasma gondii*, and *Anisakis marina*. Trichinosis has long been identified as a parasite that can live in swine muscle and be transferred to humans through raw or uncooked pork. Toxoplasma is a small protozoan that occurs throughout the world and has been observed in a wide range of birds and mammals. Anisakis is a roundworm parasite found only in fish. Using and maintaining adequate or recommended cooking temperatures and time will destroy parasites.

# MICROBES AFFECTING MEAT QUALITY

| MICROBE | TYPICAL CAUSE | EFFECT | CONTROL |
|---|---|---|---|
| Salmonella | Grows best in nonacid foods, transferred from farm animals and animal products to humans | Insulation from 3 to 36 hours. Digestive upsets. Symptoms may last 1 to 7 days. | Killed by pasteurization. Avoid cross contamination from raw meat to cooked food or food eaten raw. |
| *Escherichia coli* (E. coli) | Improper harvest methods, unsanitary handling of meat, improper cooking, fecal contamination | Severe abdominal illness; watery, bloody diarrhea; vomiting. Can affect kidney function and central nervous system. | Destroyed by internal temperatures of 160°F (71°C) |
| *C. jejuni* | Typically found in raw chicken because of high body temperature and pH | Diarrhea, abdominal cramps, nausea symptoms last 2 to 3 days | Avoid cross contamination between raw and cooked meat. Use good hygiene. Destroyed by pasteurization. |
| Listeria | Grows in damp areas, sewage, sludge; can survive freezing | Most vulnerable are infants, chronically ill, elderly, and pregnant women. Can cause meningitis or encephalitis. | Avoid raw milk products in meat recipes. Use good sanitation while processing meat. Avoid cross contamination of raw and cooked foods. Avoid postcooking contamination. |

## TEMPERATURE AND TIME EFFECTS ON MEAT SAFETY

Mismanagement of temperature is one of the most common reasons for outbreaks of food-borne diseases. This is closely followed by the time factor at a critical temperature where the correct temperature is either used too late or for too short a period.

Meat can generally be kept safe from harmful bacteria if stored under 40°F (4°C). Cooking prevents most microorganisms from growing but does not kill them, although some parasites can be killed if kept in a frozen state for various lengths of time. However, most microorganisms are merely dormant and can revive when thawed. If meat is thawed from a frozen state, it should be used as soon as possible and not refrozen.

To kill microorganisms with heat, you must maintain a recommended internal temperature for a minimum period of time. You will damage or kill microorganisms more effectively by reaching a given temperature and holding it for a period of time rather than reaching a higher temperature but for a shorter period.

Meat can be kept safe when it is hot or cold, but not in between. If meat is being cooked, it should pass between the temperatures of 40°F to 140°F (4°C to 60°C) in four hours or less. If it is being cooled, it should pass from 140°F to 40°F (60°C to 4°C) within the same amount of time.

Most, but not all, microorganisms are killed at 140°F (60°C). While the outside of a piece of meat may have become contaminated during your processing, the interior can be considered sterile, or nearly so, unless it has been cut into. When a piece of meat is cooked by conventional methods, except for by using a microwave oven, the outside cooks first and reaches a higher end temperature than the inside. Recent recommendations state that meat should be cooked to an internal minimal temperature of 160°F (71°C) because some microorganisms can still survive a 140°F (60°C) temperature. Poultry meat is more alkaline and should be cooked to 180°F (82°C), and if red meat is to be reheated, it should reach 165°F (74°C) for optimum safety. If you are grinding meat, be aware that it can become contaminated more easily than whole cuts because more of the meat particle surface areas are exposed and more processing and handling steps are involved.

Moisture in meat is essential for palatability but is also a medium for microbial growth. The level of moisture in fresh meat is high enough to provide spoilage organisms with an ideal environment to grow if unchecked. Research indicates that moisture levels in meat of at least 18 percent allows molds to grow. Drying meat through a smoking process typically eliminates moisture concerns.

Oxygen is needed for any living animal to survive but is not a welcome agent when processing meat. Oxygen is needed for aerobic microbes to grow. These include yeasts, molds, and many bacteria. Those that cannot grow when oxygen is present are called anaerobic. This group of microbes can be deadly because they include clostridium, which produces a toxin, and a group called *putrifiers*, which degrade proteins and produce foul-smelling gases. Preventing the growth of anaerobic microbes is essential if part of your food preservation plans includes canning.

Soon after an animal is harvested, the muscle undergoes a gradual change in pH, declining from about 7.0 to 5.5. This decline results from a loss of glycogen held within the muscle and its conversion to lactic acid. The degree of acidity or alkalinity (pH) will influence the growth of microorganisms. Most will thrive at a point that is nearly neutral—a pH of 7.0—than at any other level above or below. Although meat pH ranges from about 4.8 to 6.8, microorganisms generally grow slower at a pH of 5.0 or below. This acidity level helps preserve many sausages and acts as a flavor enhancer. Acidity levels are not a concern unless there is a long delay in processing the carcass at room temperatures.

A whole carcass has the minimum amount of exposed surface area. As large cuts are made, more area is exposed. When it is cut into smaller pieces, still more area is exposed. Finally, if the meat is ground, it exposes the most area for possible contamination. Simply put, the more meat is processed, the more it may be exposed to microorganisms. Using clean, sanitary equipment and clean table surfaces and keeping work area temperatures low while working as quickly as possible will help reduce microbial activity.

# INCREASED AWARENESS FOR FOOD SAFETY

Understanding the factors affecting meat quality is not intended to discourage those who want to process their own meat. Rather, it is meant to increase your awareness to the potential for problems resulting from mishandling or inadequately processing your food products.

Humans have been safely processing meat for their families for generations because they understood the basic principles to preserve meat products properly. You can also learn these principles by studying and understanding where problems could occur and take the necessary precautions and steps to avoid them.

# THE MEAT WE EAT

A common vocabulary relating to the different cuts of meat is helpful to avoid confusion whether you are purchasing the cuts from market, offering cuts for sale, or simply eating at a restaurant that lists various cuts on their menu. Having this commonality of language greatly reduces mistakes and specifically identifies the exact location the cuts are derived from on the body of the animal.

A standardized system of naming wholesale or primal cuts for each species has been developed by the meat industry in the United States and that is what is used in this book. This standardization provides for uniform sale and purchase of meat products as well as clarifies the terminology relating to them. Meat labeling practices adhere to these standards as a way to avoid misunderstanding and misrepresentation and to allow for the fair trade of meat cuts.

## WHOLESALE CUTS

Wholesale cuts are large subdivisions of the carcass that are traded in volume. Wholesale cuts are sometimes referred to as primal cuts because they are the first large portion of the carcass to be fabricated. *Fabrication* is the industry term for cutting the whole carcass into smaller, more manageable, and precise pieces. This breaking down of large portions into smaller ones allows for the sale of the more valuable cuts separate from the rest of the carcass. The fabrication process transforms a heavy, unwieldy, oddly shaped carcass into pieces that can be packaged and neatly stacked in freezers. These terms typically are used referring to domestic animals that are slaughtered but can be used interchangeably with wild game of the same general species. For instance, terms used for beef cattle, pigs, sheep, and poultry can also be applied to bison, rabbits, pheasants, and others.

Wholesale or primal cuts separate the carcass into three general divisions: the legs that make up the large muscles used for locomotion; the back and loin, which are composed of large support muscle systems; and the thinner body walls.

From these sections, the subprimal cuts or subdivisions are made. These can be further broken down into smaller pieces called retail cuts. They are often sold to consumers in a form that is ready to cook or eat.

## SHOULDER ARM CUTS

Arm bone

## SHOULDER BLADE CUTS (CROSS SECTIONS OF BLADE BONE)

Blade bone (near neck)

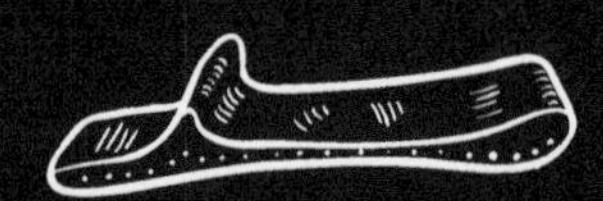

Blade bone (center cuts)

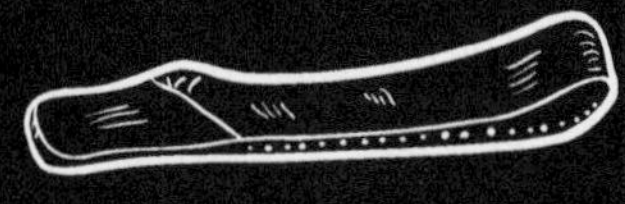

Blade bone (near rib)

## RIB CUTS

Backbone and rib bone

## HIP (SIRLOIN CUTS) (CROSS SECTIONS OF HIP BONE)

Pin bone (near short loin)

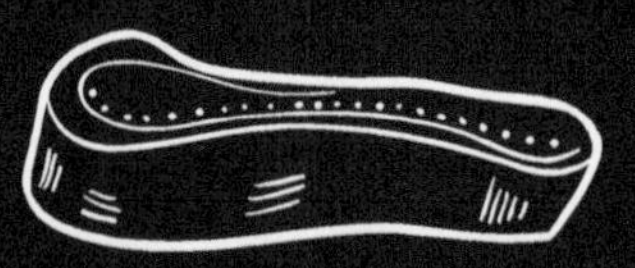

Flat bone (center cuts)

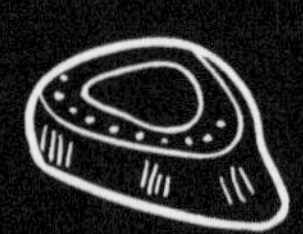

Wedge bone (near round)

## SHORT LOIN

Backbone (T-shape) T-bone

## BREAST OR BRISKET CUTS

Breast and rib bones

## LEG OR ROUND CUTS

Leg or round bone

Bone shapes in meat cuts can help identify where they are derived from in the carcass. The accompanying drawings illustrate the bone structures for the seven retail meat cuts: arm, blade, rib, short loin, sirloin, leg or round, and breast or brisket. The shapes will be similar regardless of species.

## SEVEN RETAIL CUTS

If part of the bone is included in the cut, it is called a bone-in cut and can be classified into one of seven types based on the muscle or bone shape and the size relative to it. These seven primary retail cuts include the loin, rib, leg, arm, hip, blade, and belly or plate. Familiarizing yourself with them and understanding their location on the animal or carcass will help you do a better job of cutting up the carcass.

*Loin and rib:* Two of the seven retail cuts, the loin and rib, have an eye muscle that makes up part of the back's muscle structure. The loin lies in a length-wise direction on both sides of the spinal column and is a major muscle component. Although it has use in support and movement, its location makes it less used than other major muscle groups. Less use translates into a more tender muscle texture.

The ribs are designed for the protection of the internal organs and to provide sufficient room to allow them to expand and contract as the animal eats or fasts. Ribs are connected by small amounts of muscle and connective tissue. However, their large percentage of bone to meat makes ribs less valuable as cuts but very useful for specific, lower-value cuts and soup stock.

*Legs and arm:* The legs and arm cuts have a similar cross section because of the round bones involved. However, there is a difference in the leg muscle configuration at the top, bottom, eye, and sirloin tip when compared to the arm cuts. Arm cuts are located in the front legs and contain more small muscles arranged in a different pattern.

*Hip and blade:* The hip and blade are two different cuts that have flat or irregularly shaped bones. The cuts from the hip are composed of a small number of fairly large, parallel muscles, while the blade cuts have numerous small, nonparallel muscles, much like those found in the foreleg.

*Belly or plate:* The seventh type of cut is the belly or plate. This is easily recognized by the alternating layers of fat, lean, and rib bones that make up the body wall and complete the enclosure of the internal organs. The meat taken from belly or plate cuts includes bacon, spare ribs, and beef brisket.

*Bones:* From a butchering standpoint, bones have little market value. Bones from home processing are often discarded or may be used as pet treats. However, from a culinary point of view, they can make remarkable soup stocks. The center part of the larger round bones is hollow and filled with the marrow. Bone marrow can have two different characteristics: either red or yellow. Yellow bone marrow is mostly fat, while red bone marrow is partly a fat but is interlaced with a network of blood vessels, connective tissue, and blood-forming cells. The proteins found in these cells can add to the nutritional value of sauces, soups, and other dishes.

*Color identification and size:* The color of the meat can be used for identification of species from which the cuts are derived. Beef cuts are typically large and have a cherry red color and a white, firm fat. Pork cuts are more intermediate in size and tend to have a grayish pink color. Their fat is also the softest, which makes it adaptable in making sausage and wild game cooking to prevent dryness. It is easy to understand why lamb cuts are small; they come from a much smaller animal than beef or pigs. In general, the larger the animal, the larger the cuts will be. The size of the cuts will decrease in relation to the decreasing size of the animal being butchered. Although there are cuts available in squirrels, rabbits, and other small domestic or game animals, there will only be very small pieces that can be used.

CHAPTER 2

# KNIVES AND OTHER EQUIPMENT

**YOU WILL ACHIEVE A MORE SATISFACTORY RESULT IN YOUR BUTCHERING PROCESS IF YOU USE EQUIPMENT AND KNIVES THAT ARE STURDY, SHARP, AND APPROPRIATE FOR THE TASK APPLIED. THE USE OF PROPER EQUIPMENT AND KNIVES AS WELL AS THE SAFETY ISSUES SURROUNDING BUTCHERING NEED TO BE TAKEN SERIOUSLY. INJURIES RESULTING FROM MISHANDLING ANIMALS, USING INAPPROPRIATE EQUIPMENT FOR THE TASK, AND NOT PROPERLY HANDLING KNIVES CAN BE AVOIDED BY STUDYING AND UNDERSTANDING THE IMPORTANCE OF EACH.**

Four different knives can accomplish most tasks involved in slaughtering and fabrication of animals for home harvest. These include, a 6-inch (15 cm) curved (flexible) knife, a 6-inch (15 cm) straight (stiff) knife, an 8-inch (20 cm) breaking (steak) knife, and a 10-inch (25.5 cm) breaking knife.

A large animal carcass, such as beef, has considerable weight, and the equipment you use to suspend it while working on it needs to be stable and strong enough so that it won't tip, buckle, or break while in use. Even smaller carcasses, such as deer and pigs, can have weights that challenge the equipment you might have.

Any carcass that falls to the floor or onto the ground, whether large or small, will be difficult to lift if no alternative method is available. Having to lift again a carcass that has fallen from its holding can cause delays in processing the meat, which can lead to spoilage. Also, there is potential damage from bruising of the muscles by the collapse or possible contamination of the carcass by dirt, manure, or any foreign substances it comes in contact with.

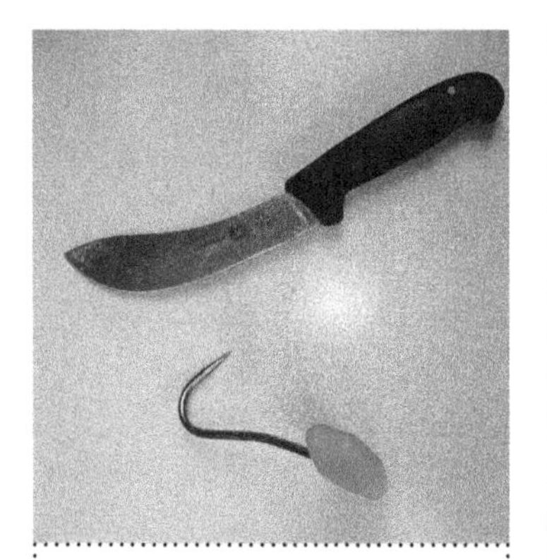

A skinning knife is slightly more curved than other knives and has a wider blade. It is used in making short, sweeping motions to separate the skin/hide from the carcass. A hook is used to pull the hide away from the carcass as it is being skinned to minimize contamination by foreign materials and dirty hands.

## SAFETY FIRST

Personal safety for you and anyone working with you is of prime importance when handling live animals, slaughtering them, and cutting up the carcasses. Being injured by live animals can have devastating consequences. Similarly, you can be injured by unstable or inappropriate butchering equipment, whether it is being used for slaughter or for food processing. Knife injuries can occur quickly and unexpectedly and, in severe cases, may be life threatening. Common sense, caution, and alertness to potential dangers will help avoid serious injury.

## THE MOST SIMPLE KNIFE RULES INCLUDE THE FOLLOWING:

- Always use a sharp knife when cutting meat.
- Never hold the knife under your arm or under a piece of meat.
- Keep knives visible.
- Always keep the knife point down.
- Always cut down toward the cutting surface and away from you.
- Never leave children unattended around knives.
- Clean knives thoroughly before storing them safely.
- Always wash knives when switching between food items.

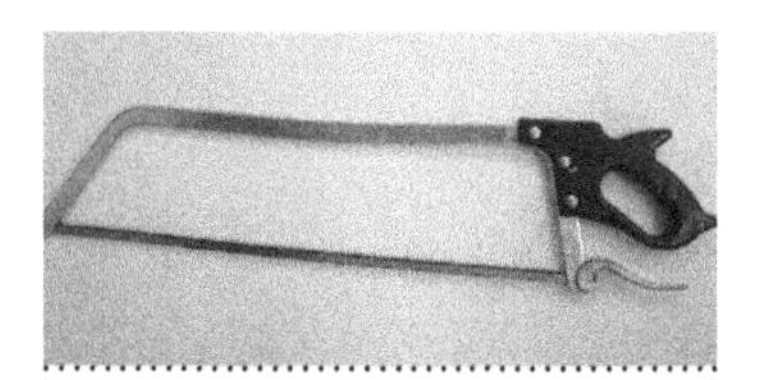

Meat saws are used to cut through bones or other areas of the carcass that may be less accessible for knife use. Most meat saws range from 12 to 25 inches (30 to 63.5 cm) in blade length.

Knives will be needed from the start of the butchering process until your last cut is made and even all the way through making jerky. The number or style of knives you use may depend on the species or size of the carcass you are working with as well as what you deem necessary to complete the work safely and satisfactorily. This may range from a small hand knife to a large, sturdy butchering knife and a variety in between. You will need knives to cut your meat into strips prior to drying for jerky. Even if you plan to use a meat slicer, chances are you will need a good, sharp knife to trim your cut of meat before slicing it.

If you need to purchase a knife, there are several things to consider. First, one knife may be able to perform multiple tasks such as slicing large or small pieces of meat, but you may find a couple of different, more specialized knives make the task easier. For example, a wide-bladed knife works well for slicing meat into thin strips for jerky making or into chunks for grinding into ground jerky. But thin-bladed, flexible knives are easier to use in deboning meat in preparation for jerky slices.

One simple rule is that sharp knives always work best. However, they also carry safety concerns when using them.

## CHOOSING KNIVES AND SAWS

An assortment of knives and saws used specifically for meat processing is available for home slaughtering and butchering. You can buy most, if not all, of the equipment used in commercial or local slaughterhouses. Purchase what you need at hardware stores, through companies on the Internet, or at stores specializing in such equipment.

When butchering, have a minimum of three types of knives available: one for sticking or cutting the throat, one for skinning, and one for eviscerating. The same holds true for cutting up the carcass and muscles. You should have a knife for larger cuts, one for boning or trimming, and one for breaking or cutting bones. You may also add to this list a saw specifically designed for use on meat.

You can purchase several knives for general use or even use knives you already have, depending on their size, condition, and intended purpose. For small animals, you may not need or want large knives or saws. You will need to use large knives and saws for big carcasses. You may want to have a separate knife for each task or you may consolidate these tasks by using only two or three different knives.

For jerky, your knife collection can range from simple, folding combination knives to a larger, sturdy straight knife. Use knives that fit your purpose, are sharp and sturdy, and are easily cleaned.

Gather together all the knives you will need before you begin butchering. Once you begin the process, you will need to work quickly and efficiently to get the animal from a live state to the freezer. Stopping to find a specific knife that is not on hand will delay this process. Take an inventory of the knives on hand and identify where they can be used during the butchering process.

Knives are typically available with wooden or plastic handles, have flexible or stiff blades, and come in many sizes and shapes. Some meat processors prefer wooden-handle knives, but these should not be cleaned in a dishwasher. Others prefer dishwasherproof plastic handles. One disadvantage to these is that they can become slippery unless dried prior to use. This problem can be mitigated by using knives with handles that have a gritty finish, which allows increased safety when they become wet or greasy. Also, the high water temperature may affect the temper of metal-bladed knives so that they will not hold their edge later when sharpened.

Buy knives that are affordable, easy to maintain, completely sanitary, and keep a sharp edge. Knives that are not sharp pose a safety hazard by not allowing you to complete the task at hand efficiently; they can slip, and more effort is required to pass the knife through the muscle or bone. If your knives are not easy to clean and kept sanitary, they may harbor harmful microorganisms that can affect the quality of the meat and possibly your health.

Identify the purposes of each knife before you begin butchering. Many knives can be interchangeable with different tasks. Always use the right size knife for the right task.

For slaughtering, a sticking knife is used for cutting an animal's throat and severing the jugular veins to obtain a good bleed before butchering the carcass. A sticking knife is generally long, thin, and has a double edge.

A boning knife has a long, straight edge for trimming and separating muscles from themselves and from the bones they are attached to. The tip of a boning knife may be ridged or flexible, allowing it to easily move around the bones. They usually range from 5 to 7 inches (13 to 18 cm) in length.

A trimming knife is a smaller, shorter version of a boning knife. It is useful for cleaning fat and tendon from small cuts and cleaning up steaks, chicken breasts, or cutting away small pieces of muscle in places that are difficult to reach.

A breaking knife is used to break down larger primal cuts into smaller pieces. It is essentially a longer version of a boning knife and is thin and curves gently up to a sharp point. A breaking knife is very useful for piercing and slicing and can be used to make primal and subprimal cuts on beef and deer.

Butcher knives are long and inflexible and are designed to allow piercing as well as cutting in a smooth linear direction. They may have either a tapered or rounded tip.

Fillet knives are long and flexible with thin blades. They are useful when trimming around bones. They are a preferred knife when skinning and cutting up fish. Fillet knives also can be used on small game animals.

A skinning knife is generally short with a dramatic curve to the blade, and it has a bulbous tip to help the blade slide easily between meat and skin without damaging either when butchering. Those used for beef are slightly more curved than the ones used for lamb or other small animals.

Cleavers are the heaviest of all butcher knives. They have a thick square blade designed to crack and split bone.

Fillet knives are long, thin, and flexible. A good fillet knife bends easily to let you cut very thin slices of fish and meat with exact precision.

Meat saws are used to cut through bones or to sever portions of large carcasses into smaller, more manageable pieces. Most meat saws are between 12 to 25 inches (30 to 63.5 cm) in length with a serrated blade. Blades should be complete, and those that have developed rust spots or have chipped or missing teeth should not be used. Any meat saw should be thoroughly washed and sanitized before use, paying particular attention to the area where the handle attaches to the metal frame.

## FOLDING KNIVES

Folding knives, as their name indicates, are those that have blades with joints that allow them to fold over, securing the edge in a protective cover—the handle. Jackknives, Swiss Army knives, and camping knives are some of the different folding knives available. They can have single or multiple blades.

Folding knives are often used during hunting because they have multiple uses, are easy and safe to carry, and are sturdy enough to accomplish quick, precise cuts. Like larger knives, they also require special care. You should keep the blades sharp and the knife clean.

Folding knives will have a locking device that keeps the blade from opening on its own. They also have a pivot that is the rotation point that allows the blade to fold into the handle. Both the locking mechanism and the pivot need to be kept clean and free of debris to prevent contamination of the meat. Use a drop of light oil at the joint, or each joint in the case of multiple blades, to create a smooth blade action while opening and closing it. As with other knives used for butchering, your folding knives should be cleaned before and after each use.

Many folding knives come with leather pouches or sheaths. When not in use, you should store the knife and leather sheath separately because leather will absorb moisture and can rust the blades.

Also, there are tanning salts and acids in leather that can rust or tarnish the steel. You can protect the leather sheaths and keep them limber by using a leather preservative or mink oil.

# ELECTRIC KNIVES

Electric knives can be used in place of standard knives. If using an electric knife, be sure it has the appropriate blade attached for the task at hand. Electric knives may be easier to use to carve or fillet different cuts of meat, particularly if handling heavy portions is a concern. Electric knives and blades will need care and maintenance like other electric equipment and should be kept away from any water source while in use.

# BLADE CONSIDERATIONS

Regardless of the different kinds of knives you use, you will want ones that have high-carbon steel blades; usually most reliable ones are at about 0.5 percent carbon. If the blade is made of too little carbon, it will be soft and the edge of the knife may bend over. If it is too high in carbon, it will generally be too hard and will be more difficult to sharpen. Try to find one that is hard enough to hold an edge but soft enough for easy resharpening at home when it becomes dull. This puts the blade with a Rockwell hardness rating of between 57 and 60.

Many knives sold through commercial outlets today are made to hold their edge or their sharpness for long periods of time and use. Older knives may not have those characteristics but may be very usable if correctly sharpened. Even high-quality knives will dull after a period of use and need sharpening. You may have them sharpened by someone specializing in blades or you may sharpen them yourself.

# SHARPENING KNIVES

In what at first appears a contradiction in terms, a sharp knife is the safest knife to use. This is because sharp knives cut more easily than dull knives, making the cut easier and safer. Less effort and pressure is required to cut through the meat with a sharp knife than a dull one, thereby reducing the risk of the blade slipping off what you're cutting unexpectedly and causing injury.

If you want to sharpen knives yourself, there are three basic steps in sharpening knives: grinding, honing, and steeling. Each is a different technique, although they may seem the same to most beginners, and each can be used depending on the condition of the knife.

## GRINDING

Grinding gives the blade the thinness and will remove part of the blade. Because of this, you will need to be cautious with any grinding so that you do not lose more of the blade than intended. Some knives need to be ground before they can be honed or sharpened. Purchased knives will come with a properly beveled blade.

Grinding produces a beveled or angled edge on the blade. In most cases, grinding is not used for sharpening, only for creating a proper angle that can then be honed to sharpen it. One of the easiest ways to grind an edge is to use a round stone that spins to grind the blade. These can be hand-turned, foot-pedaled, or electric-driven while the knife is held stationary against the stone.

Some professional knife sharpeners advise against using a power-driven grinding wheel because of the potential of creating too much heat from the friction of the wheel, causing it to burn the temper on the blade. However, if sharpened slowly in steps, you can avoid most problems with heat generated from wheels.

The purpose of the grinding process is to make one side of the blade meet the other side while pushing up a small curl of metal called a burr. If you stop grinding before the burr is formed, your knife will not be as sharp as it could be. If you grind too much, you lose any burr. As you are grinding, always check both sides of the blade all along its length. The burr tends to form quickly at the base of the blade but takes a little longer at the tip. To have fully ground one side, you must feel a burr running all the way from the heel of the blade near the handle to the tip.

Hunting knives include general-purpose types, such as a folding drop-point (top left) and a folding clip-point (at bottom). The tip of a clip-point is more acute and curves up higher than that of a drop-point. Special-purpose types include a folding bird knife (top right), with a hook for field-dressing birds.

A folding combination knife (top) with a blunt-tip blade used for slitting abdomens without puncturing intestines, a clip-point blade, and a saw for cutting through breastbones and pelvic bones of big game is another special-purpose knife. A big-game skinning knife (bottom) whose blade has a blunt tip to avoid punching holes in the hide is also useful in the field.

## STEPS TO HONE A KNIFE BLADE

- Wet the stone with oil or water and place securely on a flat surface.
- Hold the knife handle. Place the end of the knife blade nearest the handle near the edge of the stone closest to you.
- Tilt the blade so the bevel lies flat on the surface, making a 20-degree angle.
- Place your fingertips on the flat side of the blade near the back, unsharpened edge.
- Use your fingertips to apply the pressure on the blade.
- With a sweeping motion, draw the knife across the stone in one direction, then turn it and draw it in the opposite direction.

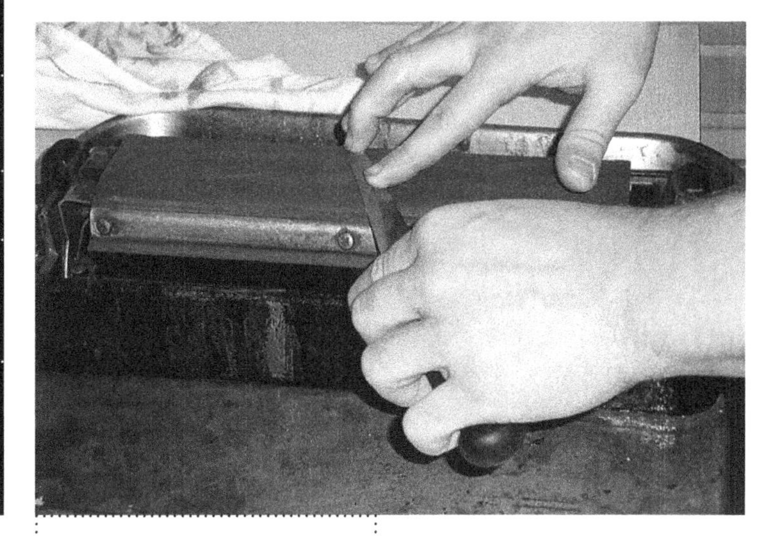

## HONING

Honing sharpens the beveled edge. You will need a stone with a finer surface than a grinding stone.
In honing, the stone remains stationary. It is important to keep the honing stone from moving while applying the blade pressure. Putting it in a wood base or attaching the stone to a table with clamps will help.

Sharpening your knife before each use will make cutting up a carcass much easier.
A sharpening and honing stone can be part of one unit, and the process for each can be changed by turning the stone over.

## STEELING

After honing, you will need to steel the blade. Steeling makes the edge perfectly straight by removing any burrs so that they do not roll over on themselves, which can cause tearing of meat when cutting, rather than slicing. A steel will realign the edge of the knife, forcing any rolled over spots back into line and making it useable again.

Knife steels come in a variety of sizes and shapes including round steels, oval steels, grooved steels, and several others not typically used in homes. A coarse steel texture will create more tiny points of contact with the edge of the blade, causing a more aggressive abrasion. You will need to be careful in not applying too much pressure so that an uneven surface is created.

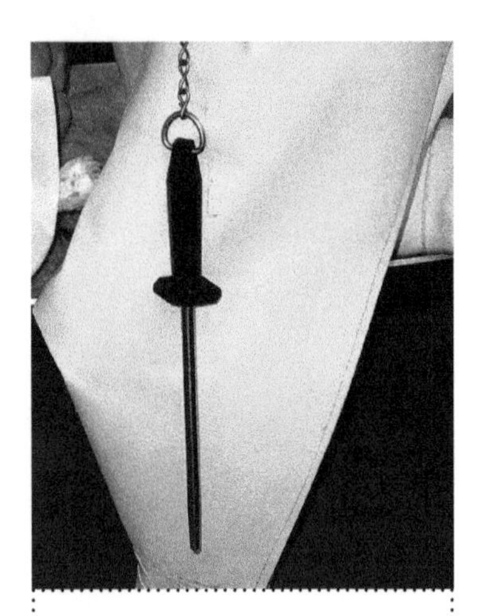

Keeping a round steel close at hand during the slaughter and fabrication will allow you to maintain the edge on your knife for easier cutting. It can be suspended from your waist by a chain.

A round steel is generally 10 to 12 inches (25.5 to 30 cm) long and can be held in one hand or placed in a vertical position with the handle up and the tip resting on a folded towel to keep it from slipping. By using this position, you will be able to place the knife edge against the steel with the blade held perpendicular at a 90-degree angle. Rotate your wrist to reduce the angle by half—45 degrees—and then rotate it again by half to about 22.5 degrees and then slightly more to a desired point at approximately 20 degrees. In general, you want to steel at a slightly steeper angle than the edge bevel of the knife.

The best result of your steeling action occurs when you lock your wrist and stroke the knife from heel to tip by moving your shoulder and slowly dropping your forearm. By locking your wrist and elbow, you will keep a stable angle from top to bottom. This is the key to maintaining a consistent angle all the way through the stroke. Standard steels do not remove metal, but only realign the cutting edge. One advantage of this method is that you won't have to apply much pressure to realign the edge. Steeling keeps the edge straight and honing sharpens it.

With a properly sharpened knife you "cut" through the carcass rather than "push" through the meat, which is often the case with dull knives. If you learn to sharpen knives correctly, it will save wear on them later. If you are unsure of your ability to sharpen knives or prefer not to, there are professional sharpening businesses that may be able to help you.

A quick steeling of your knife with a steel sharpener will keep the blade edge perfectly straight and in top condition for cutting. During a butchering session steel your knife frequently. Hold the base of the blade against the steel at the angle at which it was originally sharpened. Draw the knife toward you in an arc from base to tip. Repeat on other side. Alternate sides until the blade is sharp.

Keep your knives sharp, clean, and dry, and avoid storing them in places where they can get nicked and damaged by other objects. Even small nicks or scratches can dull the sharpest knives.

Testing a sharpened knife should be done with paper rather than your fingers. Avoid running your finger across a newly sharpened edge to test it. A better and safer method is to cut a single piece of paper while holding it loosely between two fingers.

Many different types of knives can be used for cutting meat. Regardless of the ones you select, make sure they are sturdy, easy to clean, and sharp.

A suitably sharp knife will allow you to cut through the paper with little motion.

Remember there is an inherent danger to handling, using, and sharpening knives. Knife safety, particularly during sharpening, is a matter of common sense. If you go slowly, pay attention, and stay focused, you should have little trouble. Always keep knives out of the reach of young children.

## KNIFE AND SAW CARE

You should clean your knives before and after each use to keep them in the best condition and to promote food safety. Use mild soapy water and clean by hand. A dishwasher's hot temperatures may affect the temper of the blade so it will not hold its edge later when sharpened. Also, the water jets in the dishwasher can toss your knives about and cause nicks in the blades.

When cleaning knives, you should pay close attention to the area where the blade attaches to the handle. This is the most likely area where meat or blood residue will remain after cutting and is an ideal habitat for microorganisms to grow. A thorough washing before, after, and in between cutting will maintain cleanliness.

Washing meat saws will require more attention because of the teeth on the blade. They can be cleaned with mild soapy water like knives but should never be washed in a dishwasher. Pay close attention to cleaning the teeth and the connecting joints where the blade attaches to the frame. Most meat saws have the ability to be dismantled for washing.

## STORING KNIVES

Knives can be useful for years if stored properly. Keep knives or meat saws in an area that is cool with low humidity. Avoid storage areas with a high relative humidity or that have a great shift in temperature, such as attics or basements that are not insulated or heated. Large variations in temperature and humidity can cause condensation and moisture to come in contact with knives that are left exposed.

Using a silica gel or other drying agent will help keep knives dry if you live in a humid area. Although tarnishing or oxidation is a normal part of high carbon steel knives and cannot be entirely avoided, using a gel or drying agent helps protect the knife from rust. Its residue will appear as a blue-gray hue rather than red rust tones. You can protect the blades by applying drops of any quality oil or silicon treatment with a soft cotton cloth or by removing moisture with the cloth.

If your butchering knives are to be stored for long periods between use, you should check them periodically for reddish spots that may show early signs of tarnish or oxidation—the initial rust stages. If this is present, you should clean the blades before further use. Stainless-steel blades are not rustproof, although most are rust and stain resistant. You can remove any stains or tarnish by using a standard metal cleaner or polish.

## CUTTING SURFACES

The cutting surface will have a major impact on knife blades. Always use a cutting surface that will allow you to get the most out of the knives' sharpness. Cutting surfaces or cutting boards should be made of material that is easy to clean and fairly soft. Natural wood or synthetic materials, such as soft polyethylene, are good cutting surfaces. Avoid using glass, ceramic, metal, marble, or any other hard surface material for cutting meat because these can have a damaging effect on knife blades and edges.

Cutting surfaces can provide an ideal area for cross-contamination of food products, which is a major food safety concern. Bacteria transferred from knives to cutting surfaces or cutting boards to other foods can lead to food poisoning. Always clean and sanitize the surface you use for cutting meat before and after each use.

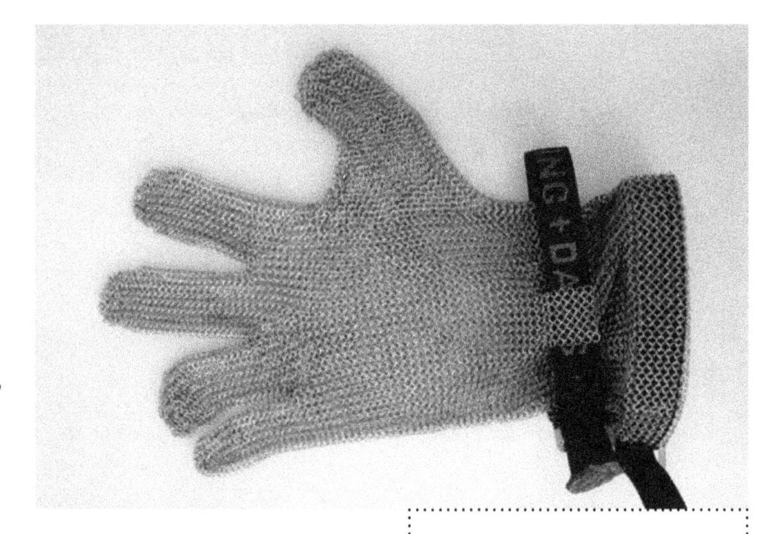

A heavy mesh butcher's cutting glove is worn on the off-knife-holding hand and is designed to protect your hands against cuts, slashes, and punctures from your knives. Always wash and disinfect the glove before and after

## OTHER EQUIPMENT

### GLOVES

Gloves are one of the best protections to use. Several different types of gloves will be useful and serve different purposes.

Rubber or latex gloves are essential when cutting up carcasses in the field because they will act as a barrier between you and the body fluids of wild game. However, they must be thoroughly washed and disinfected if not disposed of after each use.

Use a butchering glove when slicing meat in your home if you are unsure of your cutting technique or just prefer the extra safety measure. These gloves come in several sizes and are easy to wash. They are designed to be worn on the hand holding the meat, opposite of the hand holding a knife. Some gloves have braided stainless-steel threads woven into them. These prevent cuts and most punctures. Other types include a mesh glove that is made of solid stainless-steel rings that protect hands and fingers against cuts, slashes, and lacerations, but may not entirely stop punctures.

A mesh or butchering glove can be used in field dressing but, if used, you should still wear a rubber or latex glove underneath. A mesh glove is porous and fluids can seep through to your hand. Without a second barrier glove underneath, such as a rubber glove, the effect would be no different than working with your bare hands.

## KITCHEN SCALE

An accurate kitchen scale is necessary for weighing and measuring the correct quantities of spices, flavorings, additives, cure, or anything else that will be mixed into or onto the meat. Of course, you'll also want to weigh your meat to make sure you're following the recipe. Spices and flavorings do not weigh very much, so you may want to consider obtaining a digital scale that accurately measures minute amounts. A scale that can measure from ounces (grams) to a few pounds (2 kg) will cover most of your needs. These types of scales can range in price, but you should be able to find one that is affordable and accurate. Most standard scale models are easy to clean and should be readily available.

## APRONS

Aprons made from leather, Naugahyde, heavy canvas, or rubber can be a protection from injury or keep your clothes from becoming soiled or bloody during the slaughtering process of large animals. An apron will also keep you dry.

A heavy apron or abdominal protection made of material impenetrable from sharp knives is a good safeguard. While they may restrict some leg movement, such aprons are insurance against injury should your knife slip or you accidentally draw it toward your body.

Using a barrel to catch the blood draining from the carcass will keep your work area clean. It will also help in disposing or composting the blood. If you plan to keep the blood for sausage making, be sure the catch barrel or tub is thoroughly washed and sanitized, and free of any rust or foreign matter, before being used.

A digital scale accurately measures spices and meat quantities in jerky recipes.

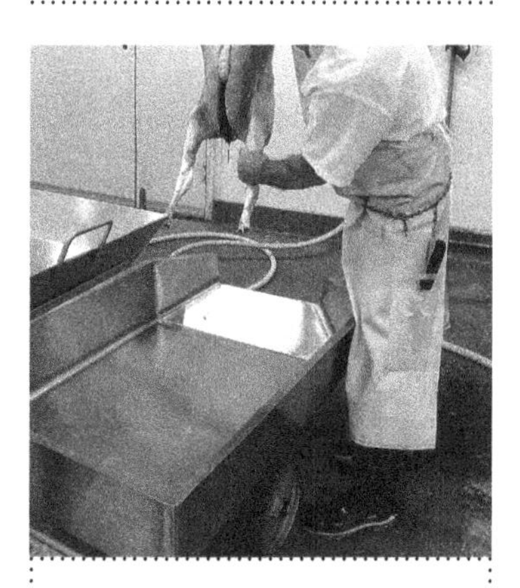

A rubber apron is easy to wash and will protect your clothing from splattering of blood and keep you dry when rinsing the carcass with water.

An abdominal apron is an essential barrier against accidental slippage of your knife. Their effectiveness in providing safety will outweigh any inconvenience in wearing one. Fully outfitted, you will be ready for cutting up the carcass.

## THERMOMETERS

Eating undercooked meat always carries health risks. The best way to monitor the correct cooking temperatures is to use a thermometer that is accurate, durable, and easy to use. There are a variety of meat thermometers available with digital models becoming more popular.

Instant-read digital meat thermometers are simple to use, provide a fast response time when checking meat, work well for thin slices of meat, and can be used to check temperatures at several spots. While they are good for checking temperatures at different points of the cooking or drying cycle, most are not designed to remain in the meat during the cooking process.

A digital thermometer model that has a probe that can remain in the meat while it cooks is also something I recommend. Most feature a probe at the end of a long cord that connects to a base unit with a digital screen that is placed on a counter or can be attached to the outside of the oven door. Because the jerky strips are about 1/8 to 1/4 inch (3 to 6 mm) in thickness, selecting a small, thin probe would work best, regardless of model.

Dial-stem thermometers can withstand high internal temperatures of dehydrators and ovens. They are more difficult to use for determining the internal meat temperature of jerky because the stem is often too thick to insert into a meat strip. However, they are useful for measuring the internal oven and dehydrator temperatures to make sure they are high enough to cook the meat completely through.

## SLICERS

Jerky making can be aided with slicers and grinders, depending on your desire for speed and precision. Electric slicers can cut meat to an exact thickness and more evenly than you can likely do by hand and knife. There are many commercial models available with working parts that are easy to disassemble, clean, and put back together. They are relatively safe to use, and their variable thickness settings allow you to alter the jerky you wish to make. Like other cutting edges, the blades must be respected. The use of a mesh glove for the hand that passes the meat along the rotating blade will help protect you from inadvertent injury.

## GRINDERS AND EXTRUDERS

Ground meat can be made into jerky, but you will need a meat grinder (see pages 177 and 199 for more on this topic). Meat grinders come in a variety of models, from hand-cranked to electric. It is sometimes preferable to use a grinder that takes trimmings and odd-shaped pieces as they may be difficult to slice with a knife or meat slicer. Whether you use a hand-powered or electric grinder will likely depend on the amount of ground meat jerky you want to make. A grinder operated by hand is useful for small amounts while an electric grinder will save time for large quantities.

A meat extruder can be used in making jerky by forming the meat into round sticks or flat strips. It is a metal or plastic hand-held tube into which mixed ingredients of meat and additives are packed and then forced through an opening by a pressure plunger. The meat to be used is first ground to a desired consistency before flavorings and spices are mixed in. This mixture is then stuffed into the tube for dispensing. It is forced out the opposite end by the plunger and, through a special attachment or tip, can be formed into a flat strip or round stick shape to be dried. There are numerous commercial models available, ranging from hand pumps to electric varieties that can handle large batches of jerky meat.

## MEAT SLICERS

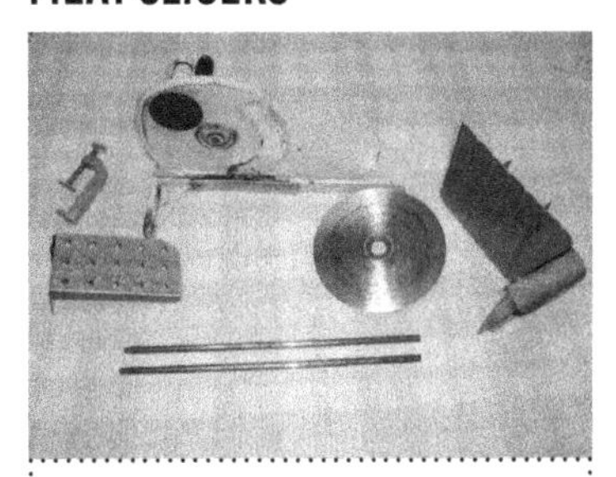

A hand-powered slicer can be taken apart for easy cleaning. Wash all parts thoroughly with hot, soapy water before and after each use.

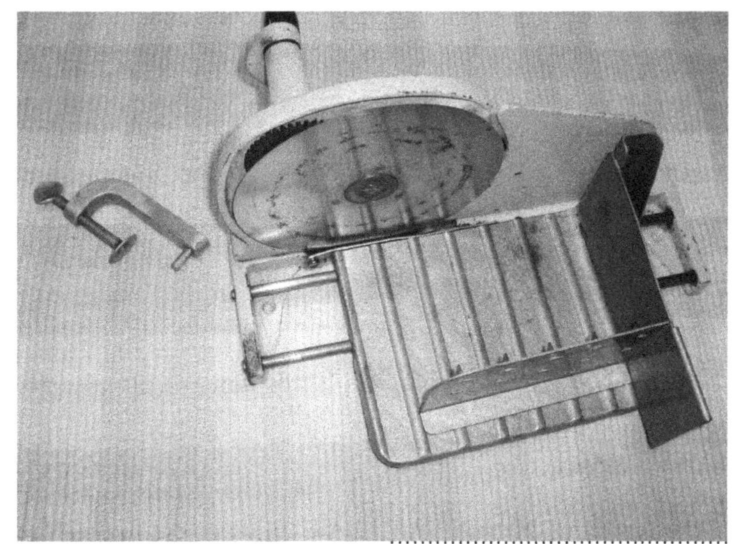

The reassembled meat slicer can be used instead of knives to create even slices. It's also good for breaking down small or odd pieces that can't be cut into strips.

## MEAT GRINDERS

A hand- or electric-powered meat grinder should be taken apart and all pieces should be thoroughly cleaned before and after each use.

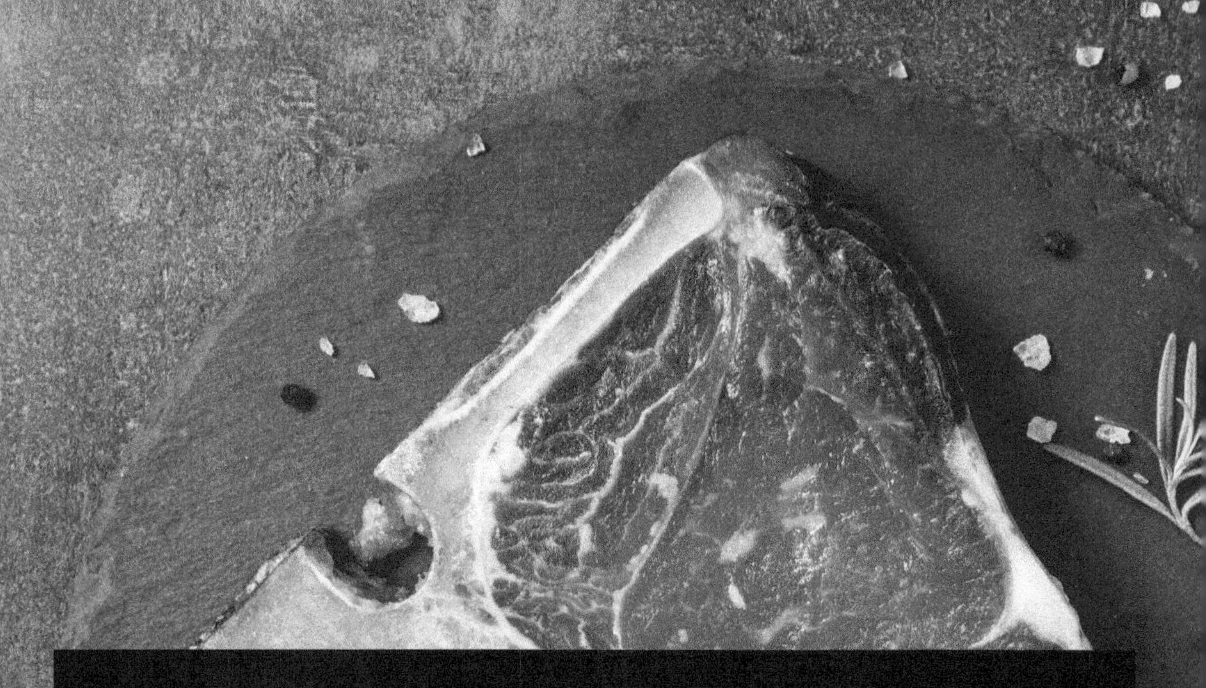

CHAPTER 3

# BEEF AND VEAL

**BEEF HAS BEEN A DIET STAPLE AND POPULAR MEAT FOR MILLENNIA BECAUSE OF ITS AVAILABILITY, NUTRITION, AND VOLUME DERIVED FROM ONE CARCASS. ITS VERSATILITY ALLOWS IT TO BE INCLUDED IN A WIDE ARRAY OF DISHES MADE FROM WHOLE CUTS, GROUND MEAT, AND STRIPS. ALSO, BEEF ADAPTS WELL TO DIFFERENT CURING METHODS SUCH AS SMOKING, CANNING, AND PICKLING.**

Select a healthy animal for home butchering, whether you raise or purchase it. Well-grown, healthy animals will yield the best carcasses both in quality and quantity of meat.

According to a 2011 USDA report, U.S. consumption of beef, pork, and chicken accounts for about 88 percent of the total meat products consumed.

Concerns about dietary fats have directed attention to bison meat because of its leanness, or higher ratio of muscle to fat when compared to conventionally raised, domesticated beef animals. This lower fat level within the carcass and muscles is perceived as a more healthful alternative.

Veal is immature beef produced from calves weighing about 200 pounds (91 kg). They are raised on diets, often indirectly dictated by consumer tastes and expectations, to produce a specific color and texture. In recent years, concern about their housing and feeding protocols has increased consumer awareness of humane veal production. As a result, growing procedures have often been altered or changed to address these concerns and minimize animal stress.

A 2011 United States Department of Agriculture (USDA) report noted that consumption of beef and veal in the United States accounts for about 34 percent of the total meat products consumed.

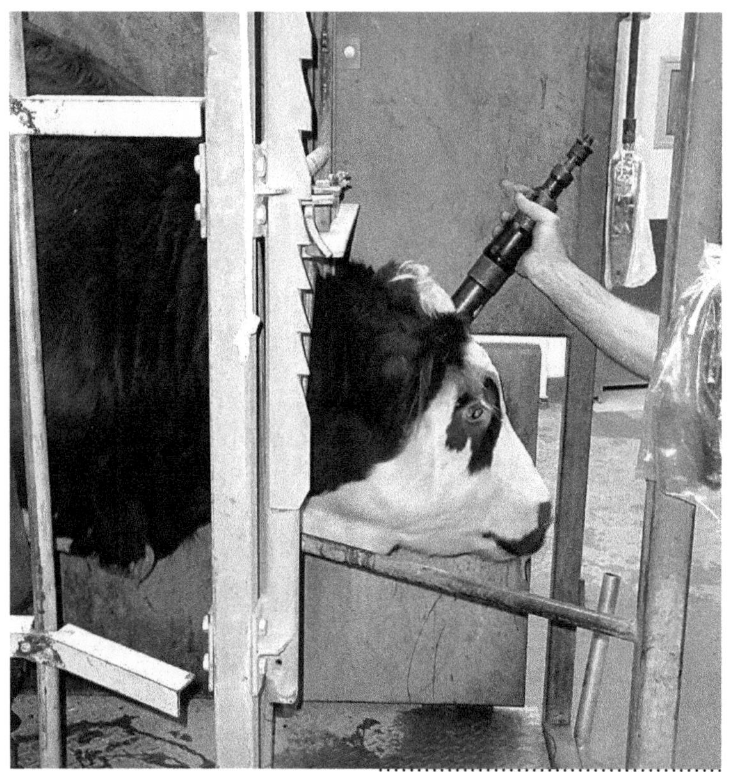

Restraining large animals is the best way to assure a clean kill and allows you to properly place the compression gun or rifle used. A stunning gun renders the animal unconsciousness so that it feels no further pain but allows for a more complete bleed because the heart is still pumping.

## HANDLING A LIVE ANIMAL

While the process for handling a carcass is very much the same in each case for beef and dairy cattle, it is the live animal that may pose a challenge. A 1,000-pound (454 kg) live animal can vary in attitude and temperament. If it must be transported, it will need time to adjust to your surroundings before you plan to butcher it.

Although a beef animal, a dairy cow, a pig, a sheep, or a goat may be more docile to work with, it is good to remember that any animal sensing a threat will react in unexpected ways. If you choose to work with a live animal, be sure you have sturdy gating and pens, a plan to quickly and safely dispatch it, and proper and safe equipment that is ready to use. Preparation for your harvest should include a thorough knowledge of the carcass, sharp and clean knives, and meat cutting saws. You must have adequate help available when needed.

You can eliminate the concerns about handling live animals by arranging the purchase of an animal and have it killed at a local meat locker. Then, you can retrieve the carcass to cut it up yourself if you have a safe, sanitary, and refrigerated means to transport it.

## CHOOSING AN ANIMAL

If you raise livestock, you will be aware of the care they need to reach a sufficient weight for harvest and which animal appears to be the healthiest for your use. If you choose to purchase a live animal from a livestock producer, make sure it is healthy in appearance. If you choose to dispatch it yourself, you will need a place to keep it until you are ready.

You should withhold any feed from your animal for at least 24 hours before you choose to harvest it. However, make certain it has full access to water so that it does not dehydrate. Cattle will lose about 3 to 4 percent of their weight if kept off feed for this period. This is called shrinkage, but it will eliminate much of the rumen contents and intestinal fecal material so that you will not have to work with it later. During this fasting period, it is very important to eliminate any excitement for the animal or unnecessary handling. Rough handling or excitement causes the blood to be forced to the outermost capillaries from which it will not drain as thoroughly as it would under normal heart action. This retained blood will lower the quality of the meat.

## PUTTING THE ANIMAL DOWN

If working with a live animal, you will be faced with the decision of how to put the animal down so that you can begin the first of the harvesting processes: the sticking of the jugular vein to facilitate a bleeding of the carcass. There are several ways to dispatch an animal, and none of them are for the faint of heart. A misapplied stun or gunshot will result in a frantic animal that will be harder to approach for a second attempt and also increase its heart rate, causing the capillary effect on the muscle and lowering the quality of your carcass.

It is best if you decide prior to harvest how you will put the animal down. A gunshot to the middle of the forehead is used by some, but it is not an effective method to create a complete bleed as possible because it causes the heart to stop beating. That makes for a slower and incomplete bleed.

You can use a power-activated compression gun if you have properly restrained the animal. They have either long or short handles and can be a penetrating or nonpenetrating type. One advantage is that they are portable and can be moved from one farm to another and from one position in your facility to another. These advantages make them comfortable to use. However, if you are inexperienced in their use, you may want to have someone who is skilled provide the service. The advantage of stunning the animal is that while it loses consciousness, its heart keeps beating, aiding in the desired blood loss after sticking.

Stun or shoot the animal in the forehead, at a point where imaginary lines from each eye to the opposite horn root, or pole, crosses. If the animal does not have horns, imagine where they would be if they did.

A major consideration about where you down the animal is how you will raise it off the floor or platform where it is standing. The weight of the animal should be a consideration in how you approach this procedure as well as the height of any ceiling present. You can use a shed or even lift the animal to an open area if you have the machine to handle it properly. Remember that once the animal is stunned or shot, you need to begin work to bleed it quickly after raising it in the air.

Your work should proceed in an area that is clean and free from dust, dirt, insects, and anything that might contaminate the carcass once it's opened. Dripping blood will quickly attract insects and flies that can lay eggs in a very short period of time. For these reasons, it is best if the sticking and evisceration is done in an enclosed area.

How you lift the animal depends on where you work. An electric winch that is firmly attached to a ceiling will work in enclosed rooms. If you choose an outdoor area, you can use a tractor with a front end lift or a skid loader that allows you to move it into place and has enough reach to keep the head from touching the ground. A 1,000-pound (454 kg) animal carcass will increase in length—as much as 7 to 8 feet (2 to 2.5 m)—as the muscles relax and stretch as it is suspended.

## EQUIPMENT NEEDED

Prior to stunning the animal, you should have all of your equipment, knives, and saws ready for immediate use. The list can be extensive as you wish, but you should have several knives available, two meat saws, a catch pan for the blood, a metal or plastic tub for the entrails, a pan for the liver, and any other items you deem necessary. A metal rod with a spiral-looped end will help with separating the trachea and esophagus. All of these items should be thoroughly washed and sanitized before they are used. A pail with soapy water and one with clean warm water should be available to wash your hands, and towels and cloths should also be handy.

## STARTING THE PROCESS

Once the animal is unconscious, you can wrap a chain around the end of the canon bone above each ankle. These bones are strong and will allow you to raise the carcass. The standard method for sticking is to make an incision, through the hide only, between the brisket and jaw. Peel the skin apart to expose the carotid arteries and jugular vein and sever them with your knife. Catch the blood in a large tub, vat, or barrel. The blood volume may vary among animals but will generally be between 6 to 8 percent of the live weight. For a 1,000-pound (454 kg) animal, this will amount to about 60 to 80 pounds (27 to 36 kg). A good stick will remove about 50 percent of the total blood in the carcass, or in this case about 30 to 40 pounds (14 to 18 kg) will fall into your catch pan, tub, or barrel.

The evisceration will be easier later if you separate the esophagus from the trachea while the carcass is suspended. If they are still attached when you try to remove the entrails, they will not come free from the thoracic cavity. It is easier to separate them at this point and makes your work less difficult later. After the bleeding is completed, you can use a metal rod that has a handle on one end and several spiral loops at the other. These loops should be threaded onto the esophagus just behind the Adam's apple and forced toward the rumen. An alternative method is used if the animal is laid on its back in a skinning cradle. Then, after the brisket is split, the esophagus is tied with string to seal it off and prevent any rumen contents from spilling out into the carcass cavity.

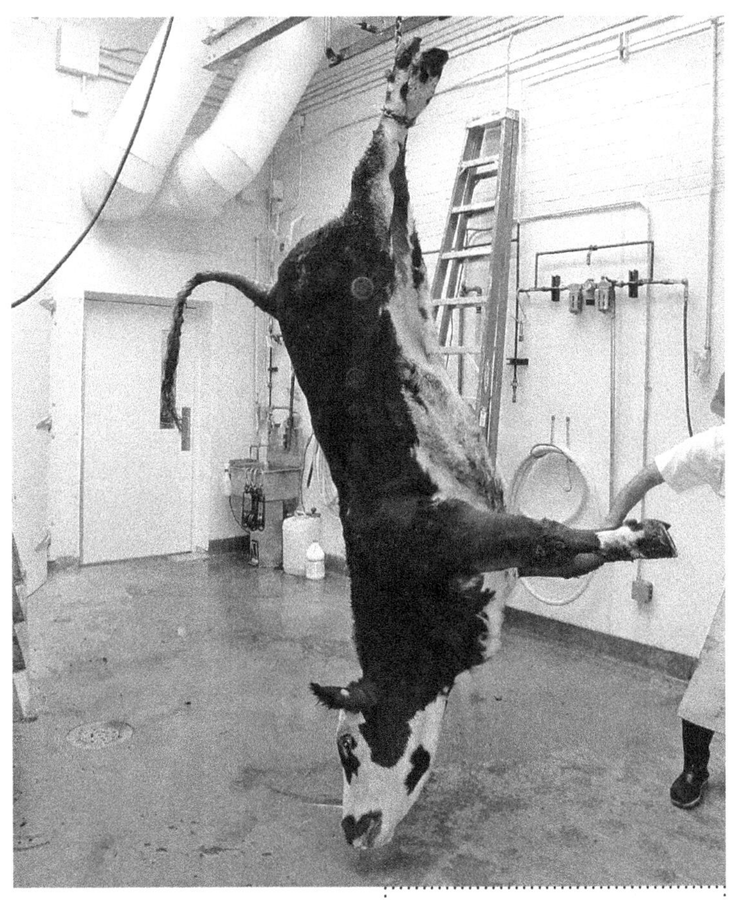

1. To raise the animal off the ground or floor, tightly chain the hind legs together between the hock and feet and lift it with a winch or loader. Be careful to avoid injury to yourself because once an animal is stunned, you may only have about 15 to 20 seconds to set the chains before the involuntary body reflexes react to the stunning and the legs begin to kick and thrash. However, they will subside within the next few minutes.

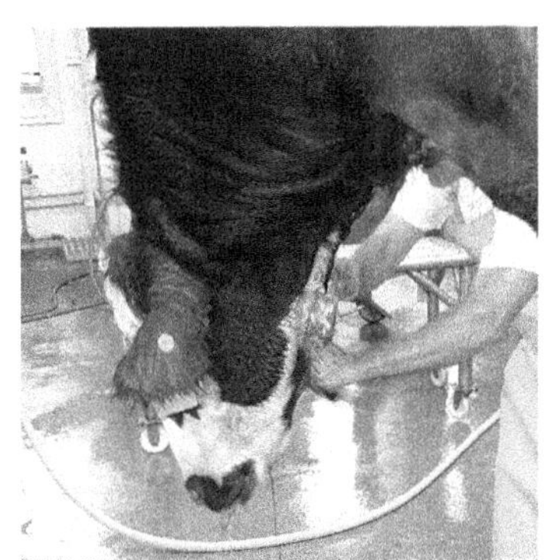

2. After the animal is raised, sever the carotid arteries and jugular vein to begin the bleeding. Make a deep incision just in front of the brisket and then down the jaw. While you are waiting for the blood to completely drain, you can begin other steps.

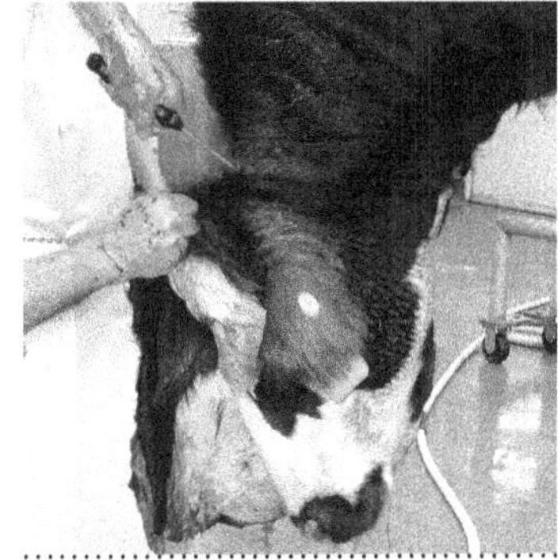

3. It will be easier to remove the head if you make a cut at the atlas joint just behind the poll or top of the head. Skin the head before you finish removing it. The atlas joint is the first neck bone and is connected to the axis joint (connected to the skull).

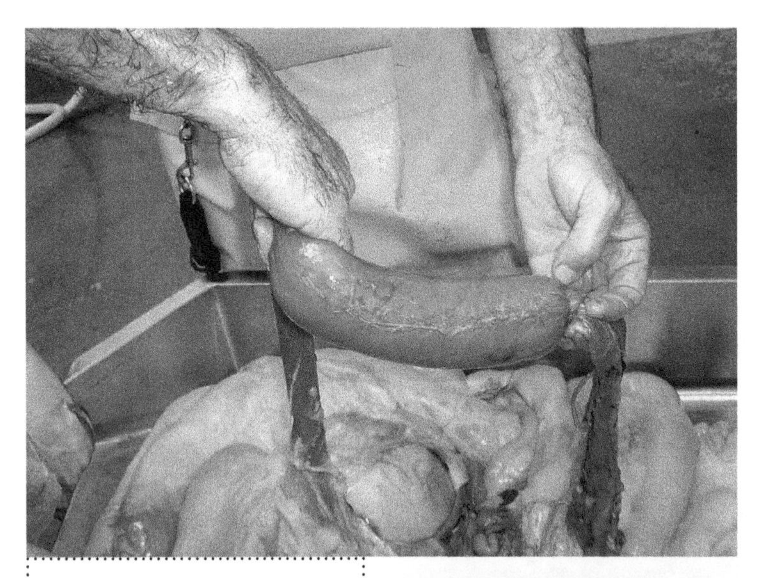

1. Tie the esophagus shut tightly with a string before cutting through it when removing the head. Cut below the string to keep all the stomach contents in the gastrointestinal tract. Without tying the esophagus closed, fecal material will spill throughout the inside body cavity and contaminate it.

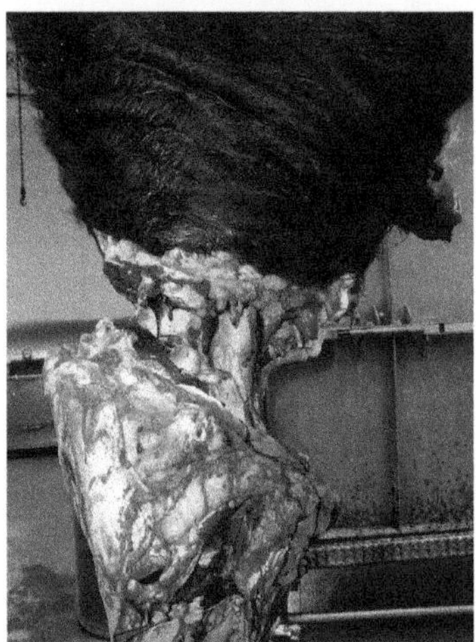

2. With the head skinned, finish cutting through the atlas joint with a breaking knife or saw. The face (cheek) muscles can be trimmed for sausage and the tongue cut out to cook as a specialty dish.

## REMOVING THE HEAD

The head should be one of the first parts removed because of its weight, to aid in bleeding, and to provide easier access to the carcass. Begin by making a cut from the poll at the top of the head down the center of the nose and down to the jaw. You can skin out one side of the face, peeling the skin back as you go, before skinning out the other side. Grasp the bottom jaw with your free hand, pulling upward so the poll bends back and cut through the Adam's apple and the atlas joint at the base of the skull. Be careful when removing the head because a 1,000-pound (454 kg) animal will have a head that weighs about 25 pounds (11.5 kg). However, it is not all waste product because you can utilize the cheek meat and the tongue.

In the past, some families made use of the brain. However, because of the development of links between bovine spongiform encephalopathy (BSE) found in infected cattle and variant Creutzfeldt-Jakob disease (CJD) in humans, you are strongly advised **not** to eat any part of the brain, spinal column, and other parts of the nervous system. Discarding the head, except for the cheek meat and tongue, and any remnants of the spinal cord is the safest route. Also, you should not feed the brain, spinal cord, blood, or other nervous system parts to other livestock or chickens. This will reduce the potential for any transference of infective agents from one animal to another.

## REMOVING LEGS

Your next step is to remove the legs to prevent possible contamination of the carcass with manure and dirt dropped from the hooves. Depending on your harvesting facility configuration, you can do this while the animal is suspended, or if you have a skinning cradle, you can lay the carcass out on its back and remove the legs.

Use the tip of your knife to open the skin, starting with a circle cut around the backside of the front leg near the dewclaw. Cut a line up the foreshank until you reach the elbow and then continue across to the midline of the brisket. Peel back the skin to expose the entire leg bone.

To remove the foreleg, cut across the shank to sever the tendon, which will release the tension on the lower part of the leg. Next, cut through the flat joint, which is about 1 inch (2.5 cm) below the knee joint. If it is too difficult to cut with your knife, use your meat saw. Then, make the same cuts on the other foreshank.

The procedure for removing the hind legs is almost identical, except you will be making your initial cut up the inside of the hind leg and across to a midline point directly below the anus. In removing the hindshank, be sure to make your cut below the point where the tendon anchors itself to the joint. This will allow you to hang the carcass by the tendons, which are strong enough to hold the weight. However, to do this, the tendons must still be attached and intact.

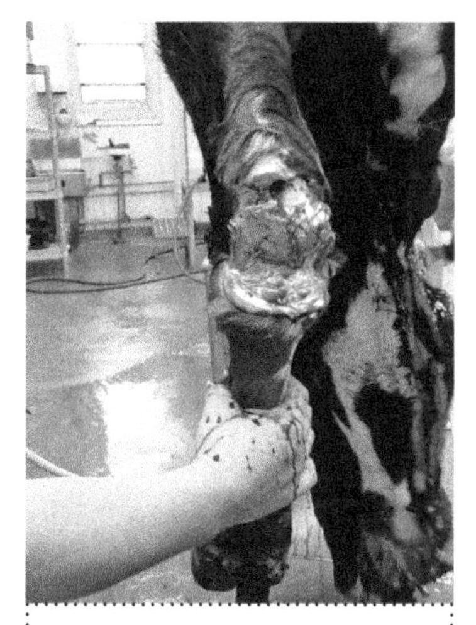

1. Remove the legs and feet first to minimize contamination of the carcass by manure or dirt attached to them. The front feet can be removed while the animal is still suspended. Make your cut about 1 inch (2.5 cm) below the knee joint, which should allow you to break it once the tendons are severed.

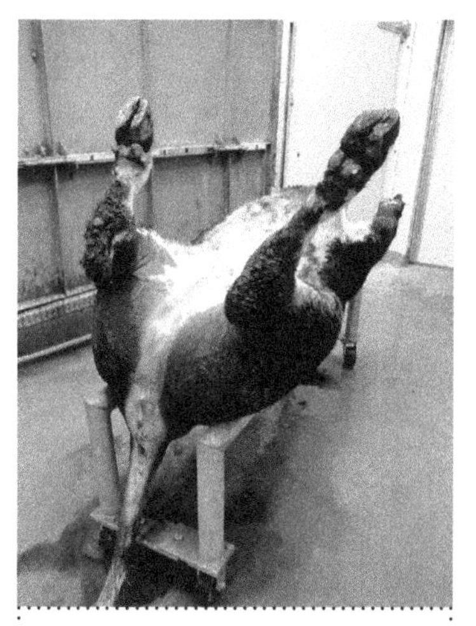

2. After the bleeding is finished and the head and front legs are removed, lower the carcass onto a sturdy trolley or platform called a cradle to begin removing the hide and the hind legs.

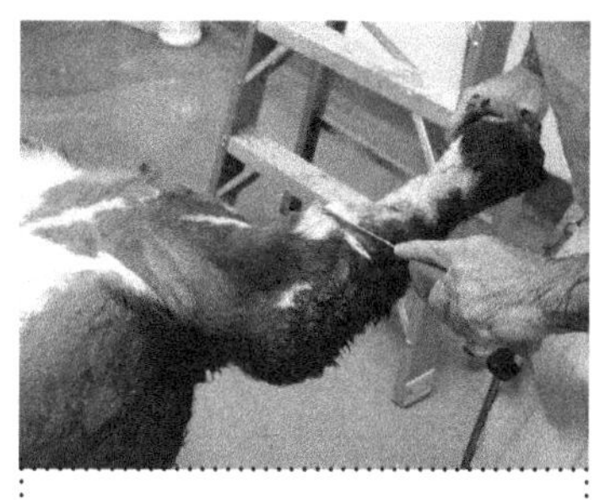

3. Remove the hind feet and legs by making cuts similar to those made in the front legs, except you will need to avoid cutting the tendons so they can be used to lift the carcass for evisceration. Make a cut at the joint located just below the hock. This will allow you to break the leg in half to finish severing it. Then, finish the other hind leg. A handsaw can also be used to remove the leg.

## REMOVING THE HIDE AND LIFTING THE CARCASS

To open the hide, you can start at either end, and this is easier if the carcass is on its back. Pull the hide upward as you make a cut from the throat to the anus, following an imaginary midline of the carcass. Pulling the hide toward you will prevent cutting into the carcass or through the abdominal wall. Next, firmly grasp the hide and use your skinning knife to make long, smooth strokes to separate and peel the skin from the carcass. Avoid unnecessary cuts in the hide if you plan to use it later for tanning.

After removing both sides of the hide as far as possible while the carcass is lying on its back, you can open the brisket. To do this, use your knife to cut through the fat and muscle covering it. When the brisket bone is exposed, you can use a saw to open it. You can separate the esophagus and trachea now unless you did it earlier when the carcass was suspended.

To lift the carcass, attach hooks to the hind leg tendons and lift so that the legs spread apart when suspended. Raise the carcass to a level that is comfortable to work with and is clear of the floor space. You can use clean chains or cables wrapped around the hind leg, but these will need to be tightly attached so they do not slip off because of the carcass's weight.

Because of the carcass's length, it will be easier to split the pelvic bone, or the aitchbone, before it is fully suspended and while still at a convenient height. It will also be easier to cut the anus loose, remove the tail, and the hide from the rump and rear quarters before lifting it. If it is a male carcass, remove the pizzle by cutting it loose from the belly and back to the pelvic junction where it originates.

Cut through the muscles and membranes at a center point in the pelvis to expose the aitchbone, using your saw to cut it in half. Loosen the anus by cutting completely around it, severing all connecting tissue. Be careful not to cut into the intestine. When the anus is loose, tightly tie the end shut with a clean cord or clean heavy string, and let it slide into the body where it can be reached from the belly cavity later. You can remove the pizzle with the anus. Remove the tail by severing the two joints where it attaches adjacent the body and cut the skin completely around its base. You should now be able to pull out the tail.

If you prefer, you can begin to split the carcass while it is in this position by using a saw to cut part way down the backbone; again, be careful not to cut into the intestines. Or you can raise the carcass until it is fully suspended and begin removing the remaining hide and start the evisceration process.

Remove the remaining hide by starting at the top and running your skinning knife down along the carcass. The weight of the hide will help separate it from the carcass. When finished, you can discard the hide or save it for tanning.

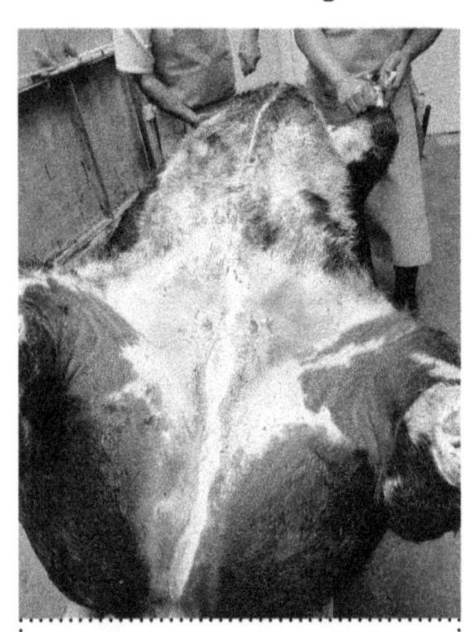

1. Begin removing the hide by making a small midline incision from the brisket to the anus. Pull the skin up and away from the body to prevent cutting into the muscle or through the abdominal wall. Rinse your knife several times to minimize contamination.

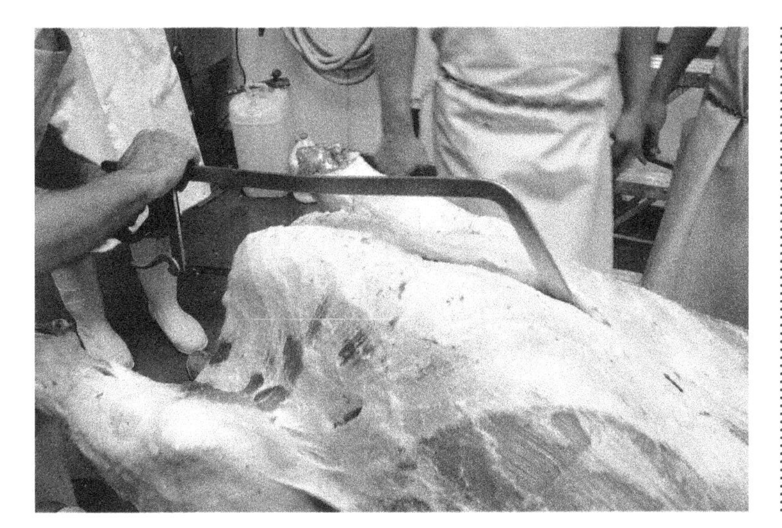

2. After removing as much of the hide as you can while the carcass is still on its back, open the brisket by cutting through the fat and muscle with a butchering knife. When the brisket bone (sternum) is exposed, use a saw to open it and expose the thoracic cavity.

3. Once the brisket is opened, pull the esophagus and trachea out. Use a weasand rod, which is placed over the windpipe and pulled through the looped rod. This separates the esophagus, which goes to the stomach, from the trachea, which connects to the lungs.

You can also strip these two apart using your hands. This separation is done so that the stomach will not be attached to the thoracic cavity during evisceration, potentially causing the esophagus to tear and result in stomach content spillage.

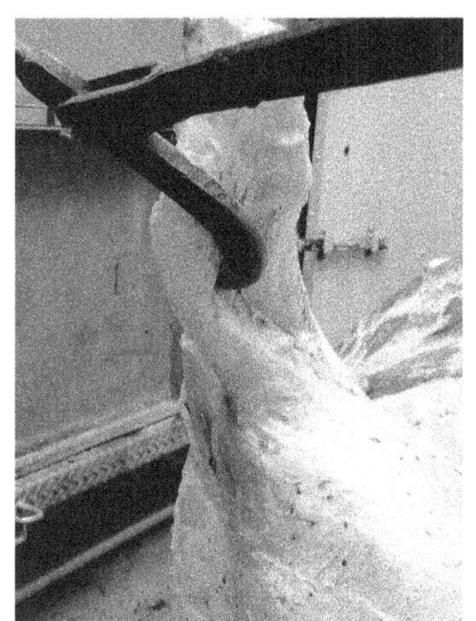

5. The carcass can be lifted by inserting the points of the gambrel between the rear shank bone and the tendon attached to it. Before lifting it, you can remove the tail by cutting through the joint closest to the last sacral vertebrae. The oxtail is often used for soup stock.

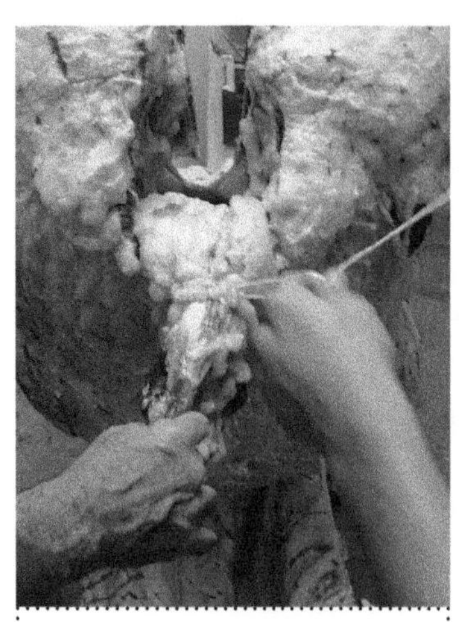

1. Several steps need to be taken before evisceration. First, make a circular cut around the anus to loosen the muscles from the pelvic bone. When free, tightly tie it shut with a heavy string or cord so fecal content will not contaminate the interior of the body cavity.

2. The aitchbone can be split with a saw either before loosening the anus or after lifting the carcass. Once the carcass is suspended, you can finish removing the hide.

## EVISCERATION

Place a tub beneath the carcass to catch the viscera after it is cut loose and to collect any blood still draining from the carcass. To open the body cavity, start at the point where you cut through to the aitch bone. Slice an opening large enough to insert your knife, handle first, into the cavity and position the blade upward and outward. This allows you to protect the intestines and rumen with your fist. You do not want the blade to cut into the intestine or rumen, as it will contaminate the carcass with fecal and rumen materials. Since you've already opened the brisket, you should make one continuous cut from the top down to the brisket opening.

As you slice down the belly, part of the viscera will spill outward but will still be held by membranes that hold the anus, intestines, liver, and bladder to the inside body cavity.

With the belly completely open, sever the fat and membranes that hold the viscera. Start at the top and cut the ureters that hold the kidneys. These can be removed later. You can loosen the liver with your hands and then sever it from the backbone with your knife. Set it in a separate pan for later inspection.

As you loosen more connective membranes, the weight of the viscera will cause it to drop outward. As it does, pull the loosened esophagus up through the diaphragm. This should allow everything to fall freely into your collecting tub.

The diaphragm separates the abdomen from the lungs, heart, and esophagus. Some people like to leave the diaphragm muscle intact and use it as hanging tenderloin. To remove the heart, lungs, and esophagus, sever the membrane and pull them out and drop them into your tub.

3. Begin evisceration with a slow and careful cut below the aitchbone and down the midline to the brisket. Avoid cutting the intestines, stomach, or internal organs with your blade. The weight of the viscera will draw it down and outward, and it can be placed in a tub once removed.

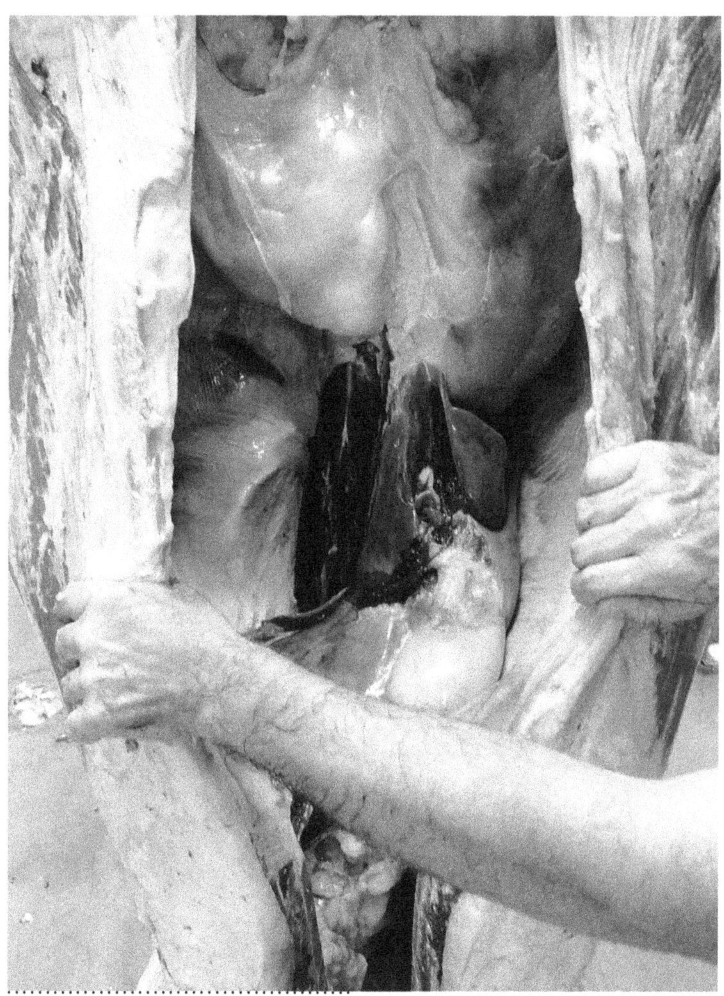

4. With the viscera removed, the diaphragm can be opened to remove the lungs, heart, and trachea. By separating the esophagus and trachea, the esophagus was removed when the viscera fell out. Once the thoracic cavity is cleaned, you can split the carcass.

## SPLITTING

With all the internal organs and intestines removed, you can now split the carcass in half. You can use your handsaw or an electric meat saw. Begin at the top and slowly make your cut in the exact center of the spine. Continue down until each half is free. Wash the carcass inside and out with cold or lukewarm water to remove any remaining blood, tissue, or foreign material. It is now ready for chilling. When you have finished with the carcass, inspect the liver and other internal organs to assess their health. A healthy-looking liver that is pink- or salmon-colored and free of lesions or dark spots will suggest a healthy animal. The liver and heart may be cooled and used later in sausage making. Generally, the intestines from cattle are too large to be very useful as casings in making sausages.

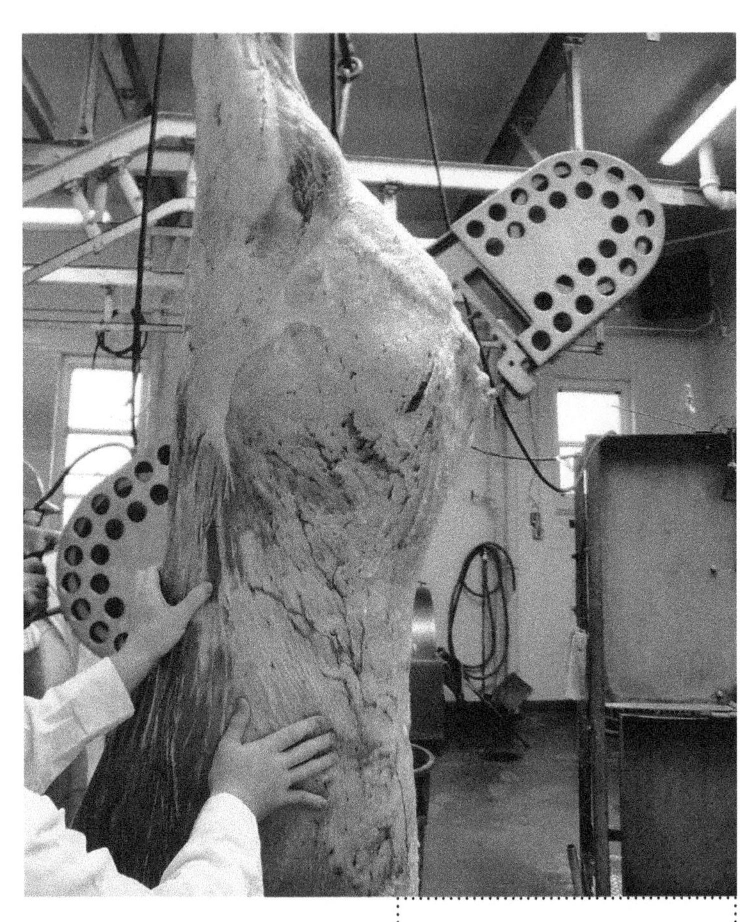

You can split a carcass using a handsaw or an electric meat saw. Start at the aitch bone and carefully cut down the middle of the spine and down the center of the spinal column. If done correctly, the backbone will be split in half, and the loin eye muscles will not be scored or cut into.

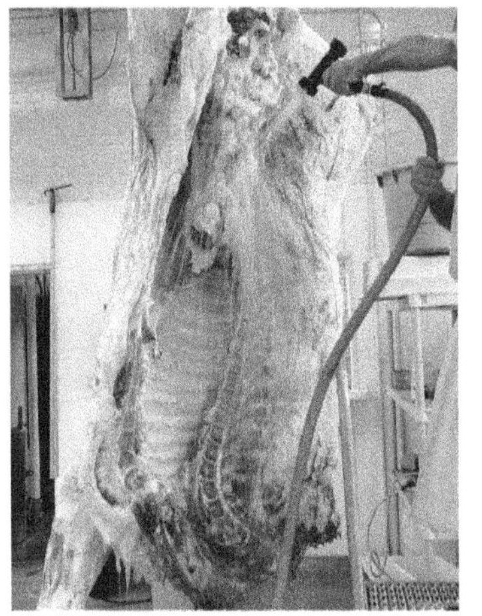

After the carcass is split in halves, wash the interior and outside with cold water. Carefully inspect the carcass and remove any remaining hair, skin, dirt, feces, blood, or other materials attached to the carcass before cooling and cutting it up. The carcass needs to be as clean as possible to minimize microbial growth on its surface.

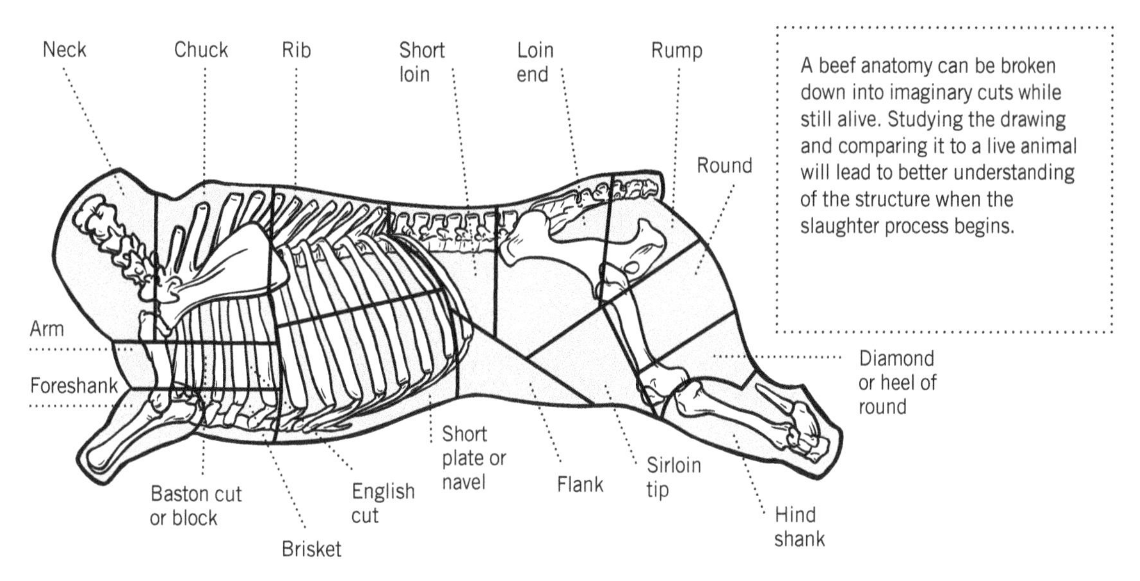


A beef anatomy can be broken down into imaginary cuts while still alive. Studying the drawing and comparing it to a live animal will lead to better understanding of the structure when the slaughter process begins.

## CHILLING

Prior to any harvest, you should decide how you are going to chill the carcass to keep it from spoiling. After the butchering process, internal temperatures of animal carcasses will generally range between 85°F to 102°F (29°C to 39°C). This body heat must be removed during the initial chilling. Meat is a perishable product and can spoil at temperatures of 40°F to 60°F (4°C to 16°C). It is important to chill the carcass for at least 24 hours to prevent the meat from spoiling. Letting it remain in temperatures of 34°F to 38°F (1°C to 3°C) will make the process of cutting up the carcass much easier as well.

Chilling a large carcass may not be feasible for a single animal, and you may have to make arrangements with a local meat service with adequate cooling facilities. You may be able to convert a large chest freezer into a cooler by setting its thermostat to a temperature just above freezing. This will approximate or mimic a still air cooler at some meat services. Depending on the size of the carcass and the cooling activity used, it may take up to 48 hours for the carcass to reach an internal temperature of 40°F (4°C) or lower. If you use a chest freezer to chill the carcass, make certain that there is space between the two sides of the freezer so that the air will completely circulate around it for even cooling. You can expect about a 2 to 3 percent loss in carcass weight during the chilling of a hot carcass immediately after slaughter. Most of this is due to loss of water.

## AGING

Aging is the process that allows the enzymes in meat to change structure in the collagen and muscle fibers that will enhance the beef flavor and increase its tenderness. Seven to eleven days is typically required to reach maximum flavor. Aging is useful for meats to be frozen but tends to decrease the shelf life of fresh meat products. There will typically be some weight loss during the aging process due to dehydration of the lean and fat. The length of time to age beef is mainly a personal preference. If unsure, it is probably better to age it for a minimum of seven days as is typically done.

There are several considerations if you decide to age beef on your own. First, it should be done in sanitary surroundings. Air should be allowed to circulate around the carcass sides completely, and avoid freezing the carcass, as that will temporarily stop the aging process. Remember that as the length of the aging time increases, so does the aged beef flavor, the tenderness, and the weight loss.

## CUTTING THE CARCASS

After the carcass has been aged, you can begin to break it down into smaller parts and pieces, which is called fabrication. Each carcass can be divided into quarters: the two forequarters and two hindquarters. Each forequarter consists of five major cuts: chuck, rib, brisket, plate, and shank. The hindquarter contains the most valuable retail cuts, including the round, loin, and flank.

Begin by dividing the forequarter and hindquarter between the twelfth and thirteenth ribs. These can be easily found by counting the exposed vertebrae rather than the individual ribs. From the rear, count off seven and a half vertebrae, reaching a point midway between the twelfth and thirteenth rib. Use a saw to cut through the vertebrae and a knife to cut through the rest.

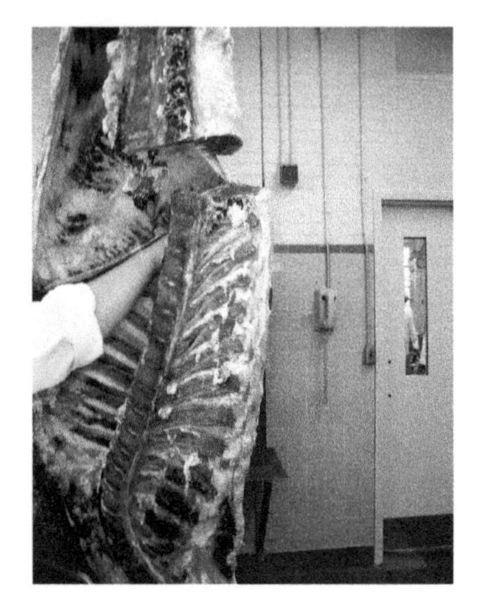

1. The carcass side is divided into a forequarter and hindquarter by making a cut between the twelfth and thirteenth ribs, counting from the anterior (front) end. Use a saw to cut through the bone, and finish the cut with a knife, splitting the side in half.

## FOREQUARTER

After splitting one side in half, separate the rib and plate from the chuck, brisket, and shank of the forequarter by making a cut between the fifth and sixth ribs, again using your saw to sever the vertebrae. Separate the plate, which is the bottom portion of the ribs, by making a horizontal cut across them starting about 10 inches (25.5 cm) below the rib eye muscle. There are two ends to this piece, one called the blade end (nearest the scapula), and the other is the loin end because it is next to the loin in the hindquarter.

The next cut should be made about 3 inches (7.5 cm) from the loin eye so that it severs the bottom portion of the ribs. Those rib ends may be made into short ribs. Next, with a saw remove the chine bones, which are located at the top of the ribs. There is very little usable meat on this cut, but it can be used for soup stock.

There is a strip of flexible but solid connective tissue called the backstrap that is still attached to the bottom portion of the ribs. This needs to be removed because it is not palatable, even with cooking.

Next, remove the blade bone and any cartilage with it. Finally, trim any outside fat off that is more than 1/4 inch (6 mm) thick. What is left is called a standing rib roast. The blade end will be larger than the loin end. If you slice the rib roast into separate pieces, those from the loin end are called rib steaks, small end. Rib steaks removed from the blade end are called rib steaks, large end. However, these will have more accessory muscles and won't be as palatable if used strictly as steaks. Another option is to remove the ribs to make a boneless rib roast or boneless rib steaks.

2. To separate the forequarter from the ribs, make a perpendicular cut to the shoulder between the fifth and sixth ribs. The shoulder, foreshank, and brisket can be set aside until needed.

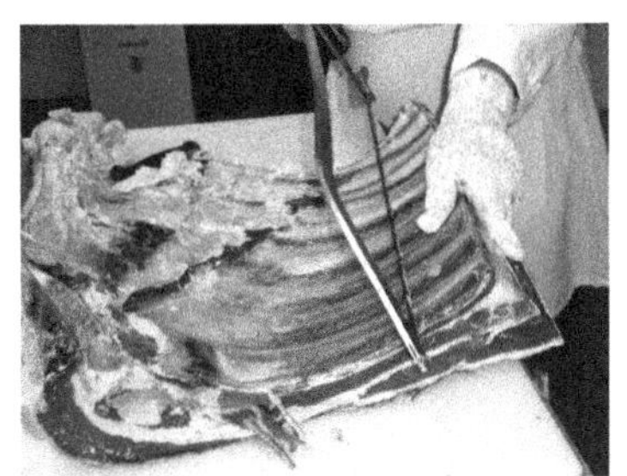

3. Trim fat from the bottom of the ribs and make lateral saw cuts about 1½ inches (4 cm) from the previous cut. These will be short ribs that can be used for soup stock or boned out for ground beef or meat trimmings for sausages.

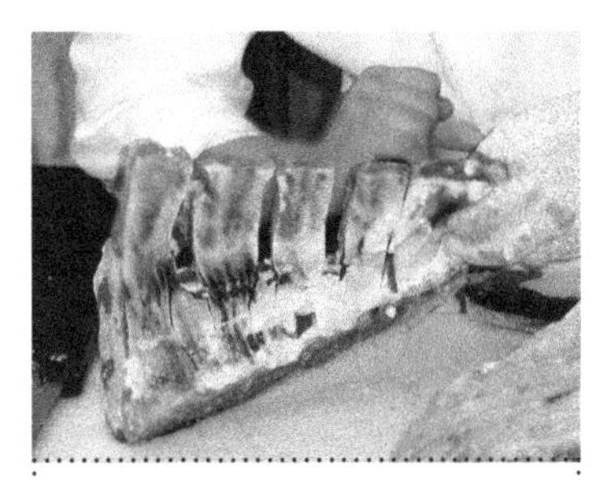

4. The ribs can be trimmed and the trim used for ground beef. It generally contains too much fat to be used in sausage making or soup stock.

5. Expect to trim and remove discolored and dried parts of the carcass, such as this piece of diaphragm attached to the plate. The longer a carcass is held in a cooler before cutting, the more that will have to be removed and discarded.

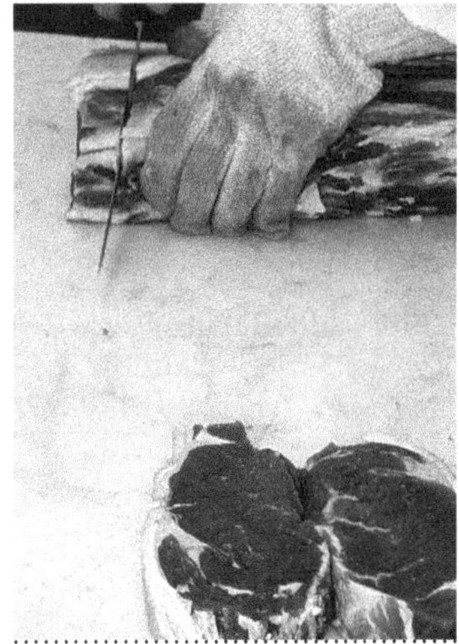

6. The ribeye roll is a valuable cut and can be sliced into 1- or 2-inch (2.5 to 5 cm) portions for grilling or braising. It can also be left intact to make boneless ribeye roasts for prime rib.

## CHUCK

After the rib and plate have been removed, the chuck, brisket, and shank remain. The chuck is the largest cut on the beef animal, and the two (right and left) will account for about 25 percent of the carcass weight. Although the chuck contains much connective tissue and is often made into roasts, there is a considerable amount of lean trim, which can be used, and several minor cuts that can be used in various dishes.

To separate the chuck from the brisket and shank, use your knife to make a cut parallel to the top side of the chuck to sever the upper part of the shoulder. Then, use your saw to cut through the rest of the shoulder bone.

As its name suggests, the square-cut chuck will have the shape of a square when you saw parallel to the arm 3 to 5 inches (7.5 to 13 cm) on the lower side of the brisket. Arm and blade roasts and steaks are made from this cut. The square-cut chuck will have fat seams even after trimming. It should be slowly cooked with moist heat for best results.

The chuck will contain some neck and rib bones that can be trimmed out by sawing across the ribs near the spine.

There is a piece of connective tissue called the backstrap that was responsible for holding the animal's head erect when it was alive. It is located on the top part of the chuck and is readily

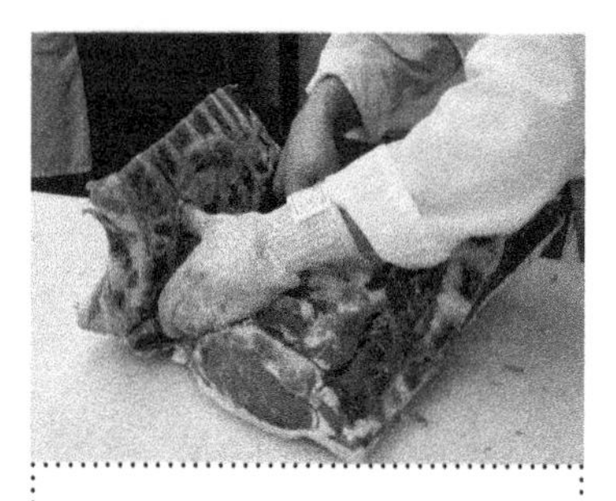

1. One subprimal cut called the ribeye roll can be made when the ribs are cut out. If the rib bones are left on, bone-in rib roasts or rib steaks can be made. By removing the bones, boneless ribeye roasts or steaks can be made. You should remove any parts of the shoulder blade that remain if making boneless cuts.

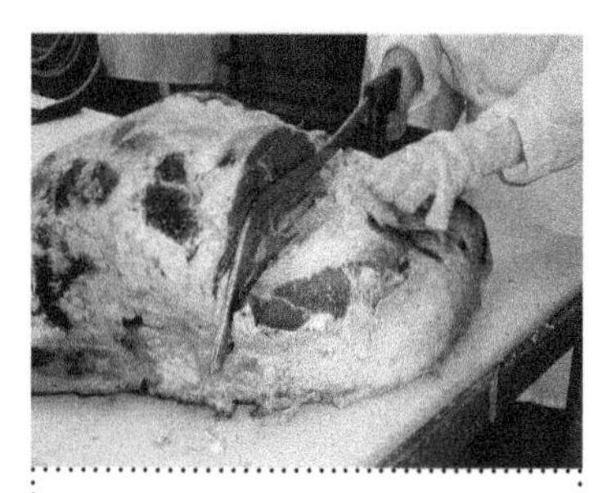

2. There is a natural seam you can cut through to remove the portion called the clod. This is typically made into roasts because it contains much connective tissue and some neck and rib bones, which can be trimmed. Practice your cutting skills on the less-valued cuts, such as the plate and brisket, before cutting the more valuable ones.

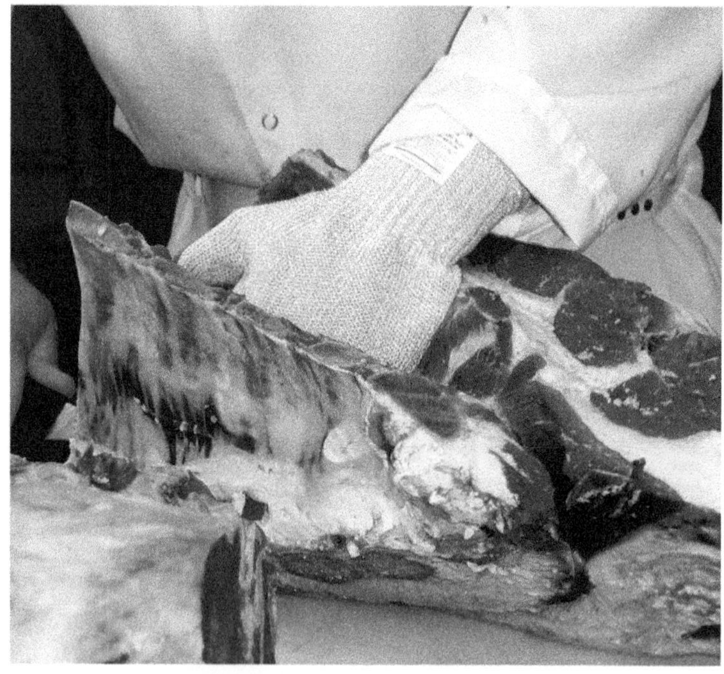

recognizable by its firm, white texture, which is impossible to make palatable. It is very similar to the backstrap you cut from the ribs, and this can also be discarded.

The chuck can be cut into several pieces. First, make two or three blade roasts by sawing across the section that had attached to the ribs. You will be cutting through the scapula or blade bone, which gives the roasts their name. A small portion of the rib eye will be in these cuts. Cut them about 1½ inch (4 cm) thick. Then, turn the chuck 90 degrees to make several cuts across the arm bone. These are called arm roasts. They will be made of fewer, but larger, muscles than the blade roasts.

After removing two or three arm roasts, you can remove several more blade roasts. As you remove these, the spine of the scapula becomes evident with the shape of a number seven. When you arrive at the neck, this can be left as a seven-bone roast, but remove the lymph node and surrounding fat deposits in it. The neck roast can be trimmed and used for ground beef because it is a low-quality cut. It can also be cut and used for soup stock.

The chuck roasts are fairly large pieces. To make them easier to work with and cook, cut them in half before packaging. You may want to trim excess fat from all these cuts prior to packaging. There are alternative methods for breaking down the chuck that you may want to study and become familiar with before beginning. Books describing different methods typically may be obtained through universities or agriculture extension offices.

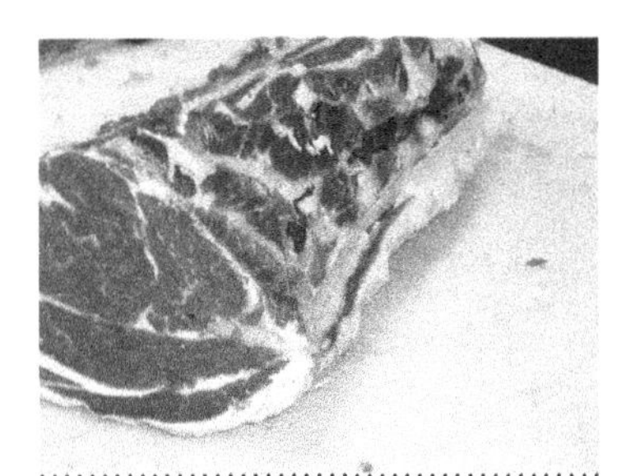

3. A strip of solid, flexible, yellowish-colored connective tissue is found in two prominent places in the carcass: the neck and the top of the back or rib section. This is called backstrap and is shown here as a yellowish strip running above the ribs. It is inedible and unpalatable, and it should be discarded. Trim these portions out of the rib section now and later when cutting up the neck.

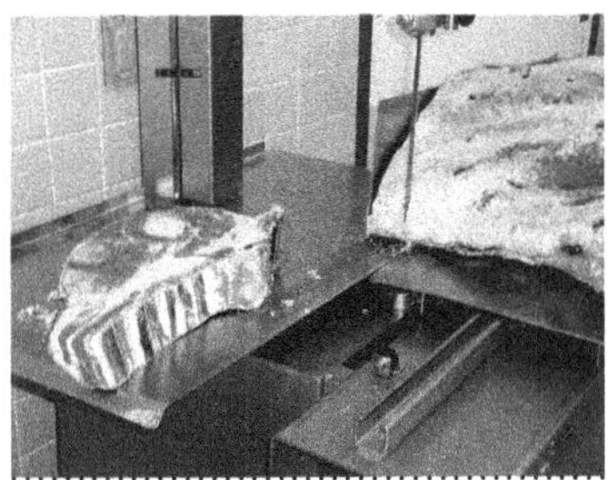

4. The chuck roll is located anterior to the ribeye roll. It can be cut into steaks, roasts, or trimmings. Typically, roasts are greater than 1½ inches (4 cm) thickness when cut and steaks are less than 1½ inches (4 cm) in thickness. A band saw can quickly cut the chuck into arm and blade roasts.

5. Animals with less finish will have less seam fat and fat between and over the outside of the muscles. Cuts with more connective tissue in the meat, such as shoulder roasts, will make them less palatable. A high degree of finish on the animal will also yield more kidney and pelvic fat.

## FORESHANK AND BRISKET

The foreshank and brisket are considered rough cuts but make up about one-quarter of the total carcass weight, about half of which can be utilized. The plate of the forequarter is the lowest part of the ribs but does not include part of the brisket. This will contain the diaphragm membrane, which should be trimmed. The plate can be used for ground beef or cut for stew meat.

Separate the foreshank from the brisket by making a cut through the natural seam that separates them. The brisket can be trimmed of all bone and used as a boneless brisket roast. Removing all the hard fat and muscles on the inside of this cut will allow you to use it for making corned beef. You can crosscut the foreshank or trim out the bone and use the lean meat for ground beef. Retain the bones for soup stock if desired.

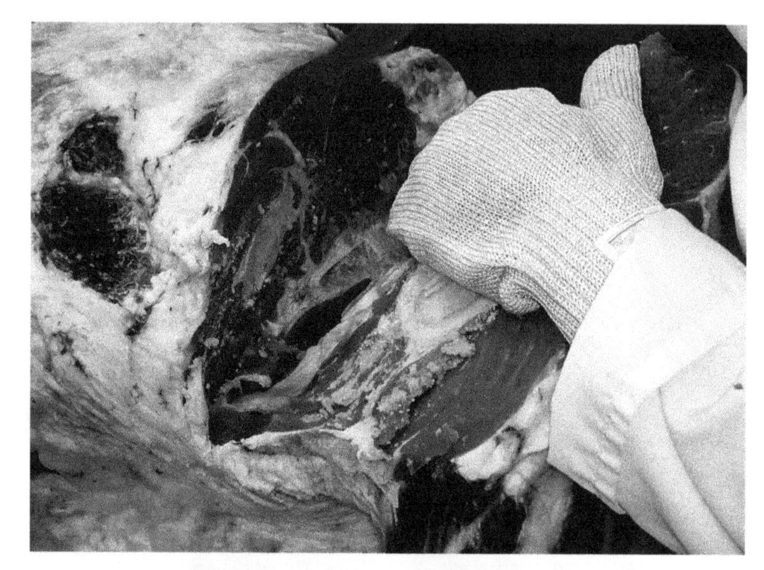

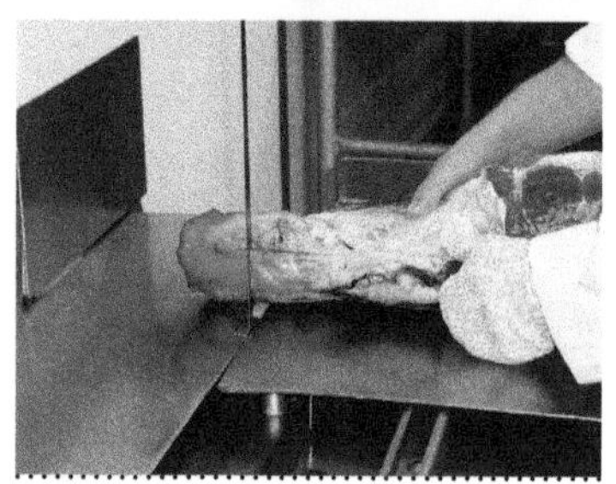

1. After the elbow is removed, there will be a choice to make. You can bore all the meat out or you can cut to make crosscut shank or soup bones. These can be cut with a knife and meat saw, or with a band saw if available. The rest can be cut into roasts.

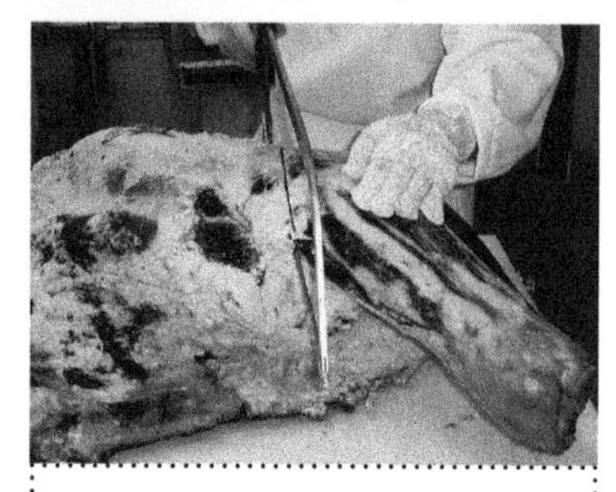

2. To remove the foreshank, make a parallel cut to it from the point of elbow 4 to 5 inches (10 to 13 cm) toward the ribs. Then, cut through the arm bone and look for the natural seam under the armpit that attaches to the foreshank and elbow and cut off the elbow with your knife.

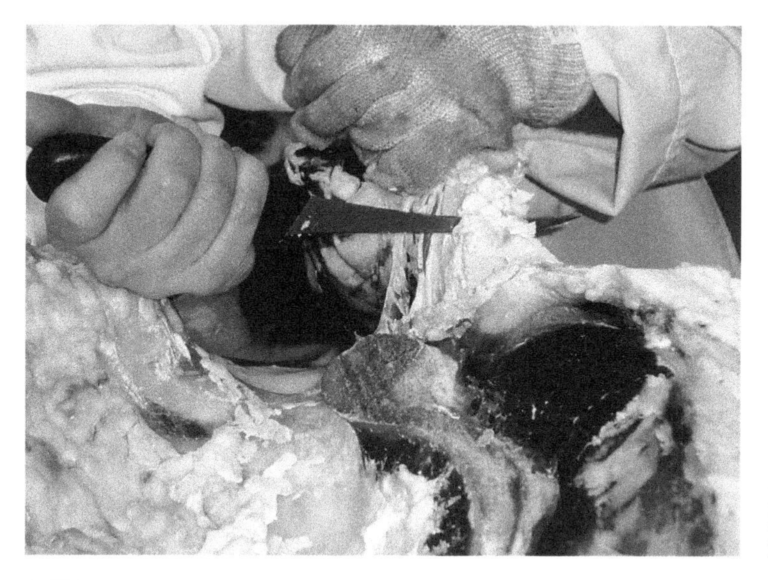

3. Four major cuts can be made from the hindquarter. These are the flank, round, shortloin, and sirloin. There are twelve ribs on the forequarter but only one on the hindquarter, the thirteenth rib. Begin by removing the flank, as there is only one per side. The tough membrane covering it can be pulled off and discarded.

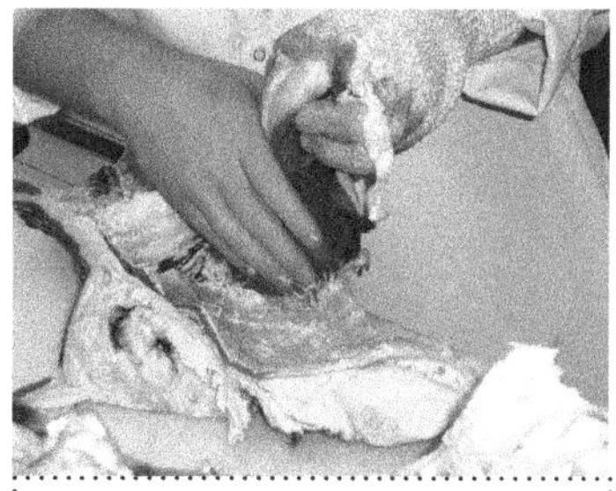

4. The flank can be easily peeled out by hand. It should then be trimmed with a knife to remove any excess fat or connective tissue.

## HINDQUARTER

The hindquarter contains three cuts that compose about half of the carcass weight: the round, loin, and flank. As with other parts of the carcass, there are several different ways to break down the hindquarter into cuts for your home use. The following is one method you can use.

One of the earliest cuts you should make is to remove the flank, which may be the easiest while the carcass is still suspended. Begin your cut by following the contour of the round—the large muscle above the hind leg, cutting toward the ribs but getting no closer than 6 inches (15 cm) to the loin. Use a saw to cut through the thirteenth rib. After this cut, you can finish the separation of the flank with a knife.

1. Remove the kidneys and kidney fat, but be careful not to damage the tenderloin area with your knife. You can leave the kidney fat intact after slaughter to reduce the dehydration of the meat, which can cause browning or brown spots on the loin, lowering its sale value and appearance. If you remove the kidney fat before cooling, don't age the carcass as long as you normally would if it was left intact.

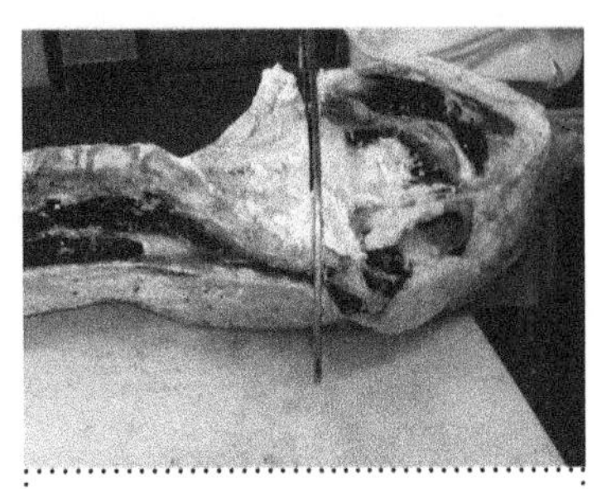

2. Remove the round by making an angle saw cut from the point immediately next to the aitchbone and through the fifth sacral vertebrae. Use your knife to finish the cut to eliminate jagged edges on the round.

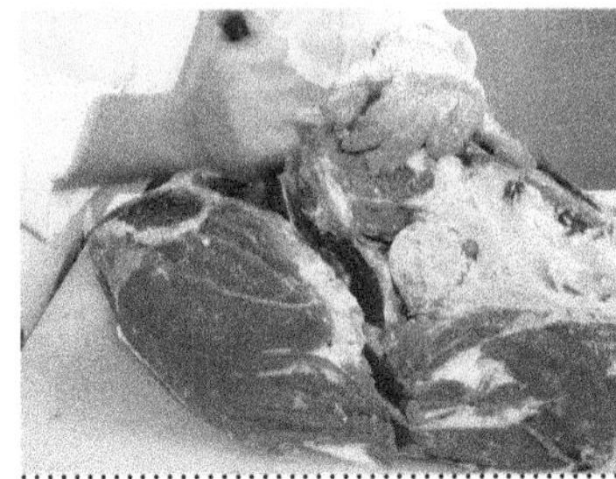

3. There is a natural seam that separates the knuckle from the top and bottom rounds. Use a knife to make this separate. You should be able to feel the femur bone as you cut along.

The flank is used mainly for trim and can be made into ground beef or used for sausage. Flank steak can be cut from each side. Because of the fat and connective tissue attached to it, you may as easily pull it from the interior surface as to trim it out. The flank steak can be broiled, marinated, or cubed.

You will not need to remove the kidneys or pelvic fat if you did this earlier during harvest. If the kidneys are still intact, you will need to remove them. They can be pulled free or trimmed but be careful with a knife so that the tenderloin muscle that lies under the fat is not cut.

## ROUND

To separate the round from the rest of the hindquarter, use a saw to begin your first cut at the rear of the aitchbone or pelvic bone. Cut just behind it and parallel to it, then saw through the large bone in the thigh called the femur. Removing the rump from the loin in this manner will provide you with two pieces without having split the sirloin tip, which may happen with other cutting options.

The round is fairly easy to cut up. The name of each cut is derived from their position when the round is laid out on a table. The top round is also called the inside round because in its natural position; it would be on the interior side of the live animal. The outside round is also called the bottom round because that is its position when it is placed on the table for cutting; it is on the bottom. The eye is located between the bottom round and the top round. The sirloin tip is that portion that is in front of the femur, or thigh bone, in the standing animal and is composed of four muscles. The top round and sirloin tip are more tender than the bottom round and the eye.

If the round has been trimmed correctly, you will see the large round thigh bone (femur) positioned in about the middle. Begin by turning the round over so that the natural seam that separates the round tip from the top round is facing upward. Use your knife to cut along this seam until the end. Then, sever the bottom round by cutting along the seam between it and the eye.

From this cut, you can trim out the round tip, which is located at the junction of the sirloin and the round. This makes an excellent roast and may be cut into steaks. The round tip cuts can be identified by the oval- or horseshoe-shaped connective line in the center of each cut.

Cut the round steaks to a width of about 1 inch (2.5 cm) after you have removed the round tip by cutting across the face of the round. These will be large pieces and can be folded over to be packaged or cut in half.

Next, remove the hind shank bone. Cut the Achilles tendon, which had held the weight of the carcass as it was suspended. Strong connective tissues in the shank anchor the muscles in the lower round. These must be severed to remove the shank bone. Tip your knife up and cut along this bone up into the round to the stifle joint. You can sever this joint with a knife and trim any remaining connective tissues; then remove the shank bone.

Turn the round over and trim out the rest of the stifle joint, making sure to cut close to the bone. The bottom round then can be separated from the top round by following the natural seam between them. You can cut steaks from the top round as they are considered more tender than other parts of the round. Steaks can also be cut from the bottom round if desired, or the meat can be used as cube steaks.

## RUMP

The rump is considered as part of the round and can make up about 4 percent of the carcass weight. To separate the rump from the loin, make your cut along a line that would connect a point on the backbone between the fifth and first vertebra of the tail and the front tip of the inside end of the femur.

Begin trimming by cutting closely on both sides of the aitchbone until it is free. The remaining cut is the boneless rump that will tend to spread because it is not connected to any bone. It can be wrapped with netting to hold its shape.

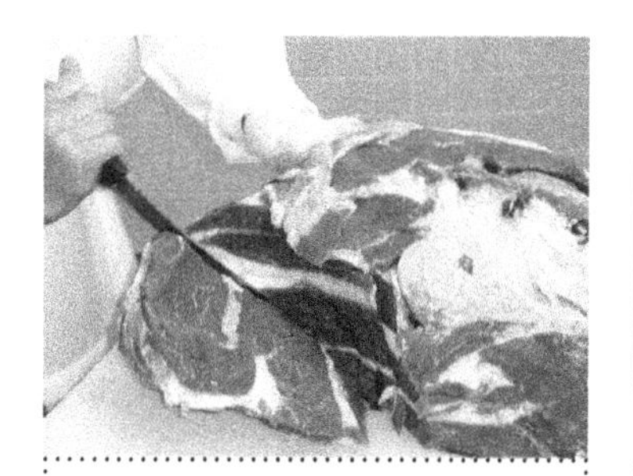

1. From the cut just made, you can trim out the round sirloin tip, which is located at the junction of the sirloin and the round. This can be cut into 1-inch (2.5 cm)-thick steaks or left as a roast.

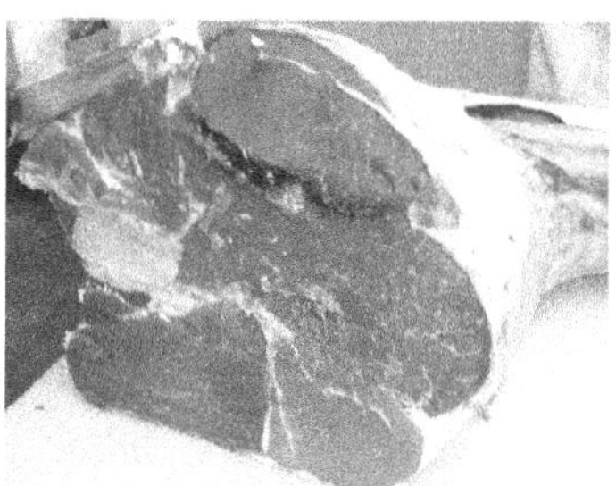

2. After the knuckle has been removed, the top and bottom rounds will be left. These will be large slices because this area of the round includes several muscle groups with different striations. You can cut round steaks or separate the top and bottom portion into boneless top round steaks/roasts and boneless bottom round steaks/roasts.

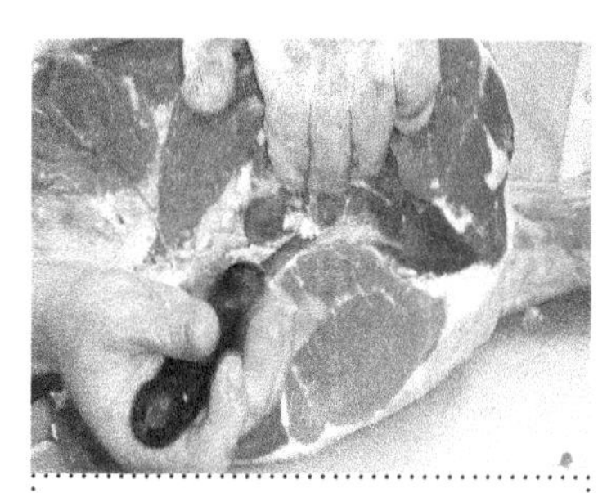

3. Begin removing the top round by inserting your knife in the natural seam between the top round (top of picture) and the bottom round.

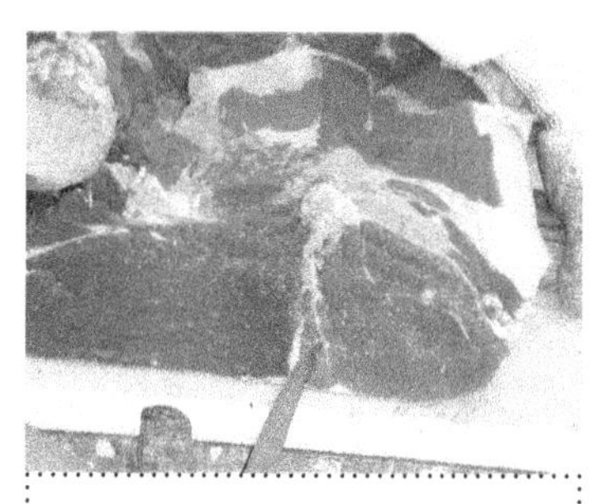

4. After removing the top round, the remaining portion of the rump is composed mainly of the bottom round (on left) and the eye of round (on right). Connective tissue and fat should be trimmed off with your knife.

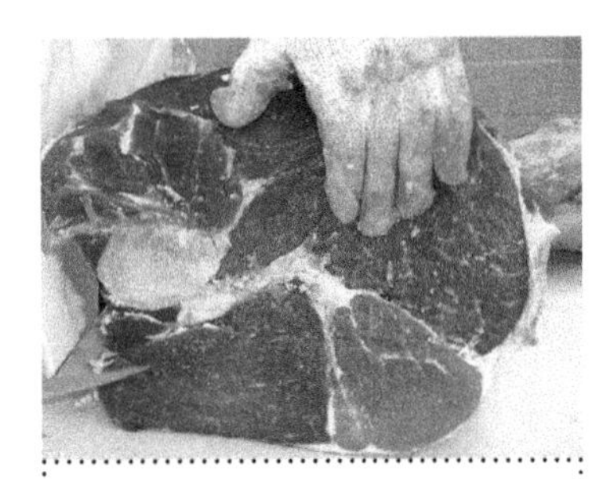

The eye of the round is a boneless cut made by separating it from the outside rounds, following the natural seams. On this view, the eye of the round is located as the bottom right triangular piece and can be separated from the other muscles by natural seams.

## LOIN

The whole loin is composed of two parts: the sirloin and shortloin. The steak yields from the sirloin are about 5 percent of the carcass weight while the shortloin will be about 7 percent.

For the loin, make your cuts between the vertebrae. If you are cutting the entire loin into steaks, there will be different sizes and shapes as you move from front to back. The round bone steak contains the most meat of all the steaks of the sirloin because it has the smallest amount of bone and few fat seams.

Porterhouse steaks come from the end of the sirloin nearest the shortloin. They are easily identified by the large size of the tenderloin. The porterhouse also has an extra muscle attached to it, which decreases in size and disappears when you arrive at the shortloin. From this point on, the steaks become T-bone steaks, identified by their characteristic letter-shaped configuration.

You can trim the tenderloin as one cut if you desire. It can then be trimmed of accessory muscles and fat. Tenderloin fillet steaks come from trimmed tenderloin. It is the most palatable meat in the beef carcass.

When you have completed the last cut on this beef side, you can begin the same process with the second side of the carcass.

The whole loin is made up of two cuts, the sirloin and shortloin. The sizes and shapes will vary as you move from the front of the loin to the rear. They can be cut into porterhouse and T-bone steaks. The difference between them is the size of the loin eye. The porterhouse has a larger tenderloin than the T-bone.

When cutting T-bone steaks, don't stack them on top of each other for very long or you will get brown spots due to oxidation or muscle exposure to air. This can be avoided by putting plastic wrapping over each piece.

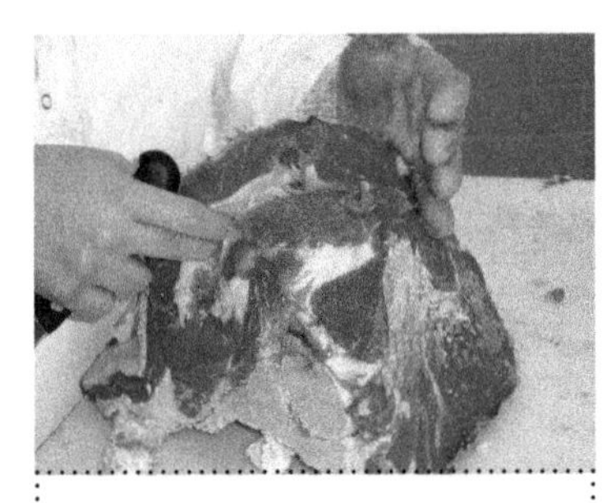

A cut called the tri-tip, or angular roast, is located below the bottom sirloin. To trim it out, follow the natural seams because it has muscle groups running in three different directions.

## VEAL

The harvest of a veal calf is very similar to that of mature cattle, except the obvious difference in animal size and subsequent carcass weight and the amount of meat procured.

Veal is identified as meat from calves of all ages and weights, from birth to 20 weeks of age. There are four classifications of veal recognized by the meat industry: (1) baby, or bob veal, that includes calves 2 to 3 days to 1 month of age; (2) vealers, 4 to 12 weeks old, 80 to 120 pounds (36 to 54.5 kg); (3) calves up to 20 weeks, 125 to 300 pounds (57 to 136 kg); and (4) nature or special-fed veal, about 20 weeks of age, 180 to 240 pounds (82 to 109 kg).

There are several important differences to recognize in harvesting a veal calf. First, there is very little fat cover on the body, and the carcass and muscles will have a higher moisture content than older animals. Second, the size of the cuts will be smaller, and some may be treated in a similar way to lamb cuts. For example, you may leave the sirloin on the leg instead removing it as a separate cut.

The meat of a veal calf will be pinkish white to grayish, or light pink in color. Veal has a mild, delicate flavor and is often served with sauces and/or spices.

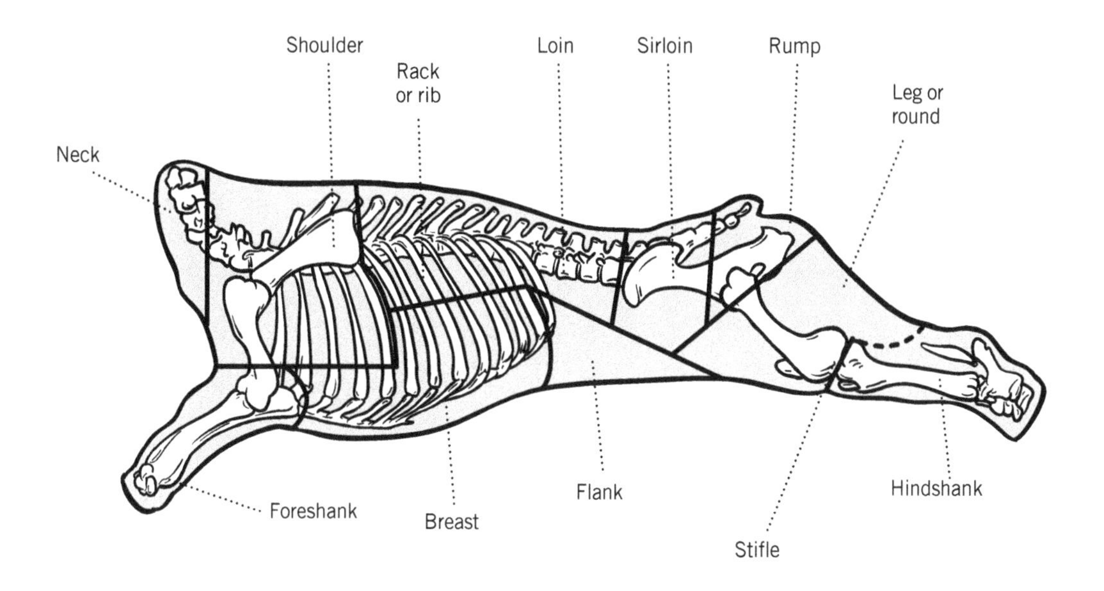


The anatomy of a veal calf is similar to a large beef animal. This drawing shows the skeletal structure in relation to some of the wholesale cuts.

CHAPTER 4

# SHEEP, LAMBS, AND GOATS

**MEAT FROM SHEEP, PARTICULARLY LAMBS UNDER ONE YEAR OF AGE, AND GOATS IS INCREASINGLY POPULAR IN RESTAURANTS, GROCERY STORES, ETHNIC FOOD OUTLETS, AND THE EVERYDAY FAMILY FOOD TABLE. THE MEAT IS HIGH IN PROTEIN AND CONJUGATED LINOLEIC ACID (CLA), A UNIQUE AND POTENT ANTIOXIDANT NATURALLY PRODUCED THROUGH PASTURE GRAZING.**

The meat from sheep up to one year of age is referred to as lamb and is usually taken from an animal that weighs between 90 and 140 pounds. (41 to 63.5 kg) Lamb is typically sold as whole or half carcasses if you decide to purchase one for cutting up yourself.

Whether you are harvesting sheep, lambs, or goats, make certain the animal is healthy and that you have the proper equipment and an appropriate place to butcher that is clean and free of dust, dirt, insects, flies, and rodents. Also have sufficient help on hand and the physical ability to carry out the work.

Lambs are typically marketed at between 90 to 140 pounds (41 to 63.5 kg) and sold as half or whole carcasses.

# FIRST CONSIDERATIONS

A well-devised butchering plan will help achieve good results. It starts a day before you plan on slaughtering the sheep, lamb, or goat. Sheep and goats have a digestive tract that has a higher percentage of the live weight than other livestock, such as cattle and pigs. This makes it important to withhold all feed for between 18 to 24 hours. Fasting your animal will allow it to empty the stomach and intestinal tract of fecal material that has the potential to contaminate the carcass during evisceration. You will still need to provide full access to water. Providing water will avoid dehydration, which can result in tissue shrinkage and difficulty in removing the pelt.

# CHOOSING AN ANIMAL

If you raised your own sheep or goats, you should be able to determine the healthiest one in your flock as a good candidate for butchering. It should also be one with the most muscling and least fat.

If you are purchasing a live sheep, goat, or a lamb, be sure to examine it first. The eyes, nose, and mouth should be clean with no watering or discharge. It should move about easily without exhibiting any lameness or limping. The presence of either may indicate an injury or other physical illness that will lower the carcass and muscle quality. Observe the animal's breathing pattern. If it is labored or fast, it may indicate lung problems or a fever. Refuse to purchase any animal if the physical signs you see do not appear as normal or you sense something might be wrong.

You may want to consider purchasing a female for butchering to avoid removing the male sex organs. Some believe there is a distinct difference in meat flavor between a female and intact male.

After slaughter, the lamb carcass will weigh roughly 50 percent its live weight. Depending on your storage capacity, you may be able to purchase a whole or half carcass for butchering.

# HANDLING

Proper handling of sheep and goats at all times is good husbandry and minimizes damage or injury to the live animal. It is particularly important during the time leading up to slaughter to reduce the chances of damaging or bruising the muscles. Bruised muscles yield a lower-quality carcass and, if severe enough, may require the bruises to be cut out of the meat, lowering your total yield. To prevent bruising, provide sufficient room for the sheep to move about and still be caught without injury. Avoid lifting it by its fleece or hair, as this will also cause bruises to the carcass. When moving your animal to the confinement area, place one hand under its jaw and the other at the dock (tail) and lead it.

Be careful of the sheep or goat if it has horns, which can be used in defense if they perceive to be in imminent danger. Sheep or goat horns have pointed tips, which can cause serious injury to anyone handling them.

The process to harvest a sheep or goat is essentially identical, so the following descriptions apply to each.

# EQUIPMENT AND TOOLS

The minimum equipment you should have available include a sharp skinning knife for removing the pelt, a table or platform on which to lay the eviscerated carcass, or a hoist to lift the sheep by the hind legs to eye or chest level. If available, you can use a cradle, which is a trough about 6 inches (15 cm) wide at the bottom with sloping sides 6 inches (15 cm) high to be used for skinning. A dripping pan to catch the blood after sticking will keep the area below the carcass from becoming messy. Using a chain-mail glove on your free hand will prevent accidental cuts.

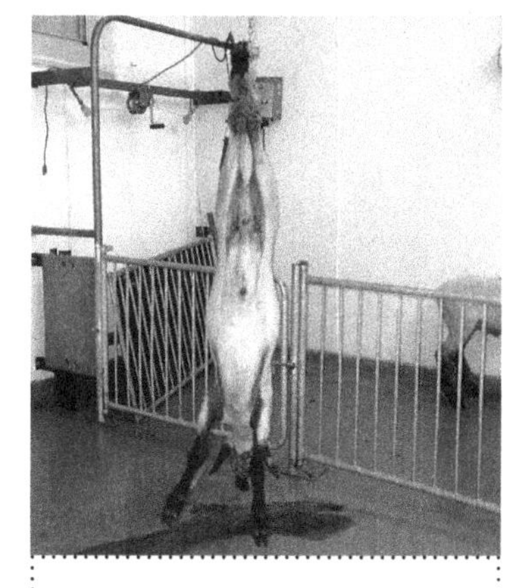

Lift the lamb by its hind legs to complete the bleed.

## CHOPS

The term *chop* originally referred to any piece of meat that was chopped off with a cleaver. These were usually only small cuts because of the difficulty of chopping huge pieces of meat off at one time. If a saw was used, the pieces were called steaks. This is why lamb cuts, except for the leg slices that are steaked, are called chops.

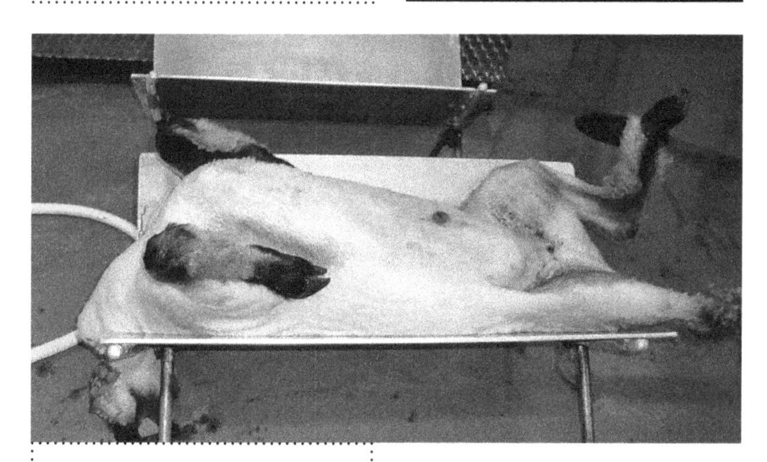

Once it is finished, place the carcass on a table or a V-shaped trough to begin removing the legs and pelt.

Stun or shoot the lamb as close as possible to the place where the imaginary lines intersect.

With restaurants, grocery stores, and ethnic food outlets offering more and more options, meat from sheep, particularly lambs under one year of age, and goats is increasingly becoming a part of the American diet.

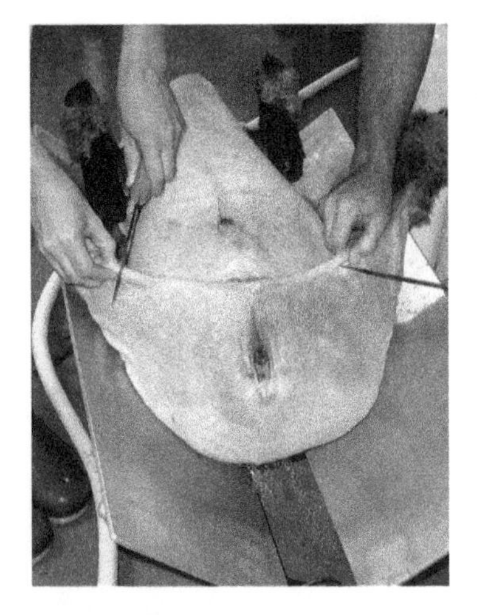

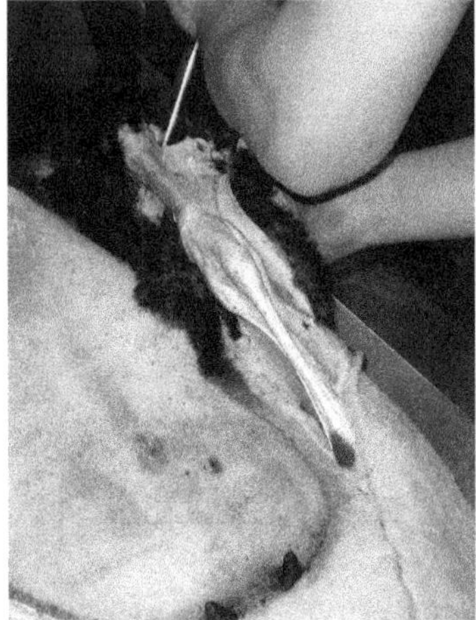

Begin removing the pelt by slicing down the rear legs from the hock to the midline of the pelvis and down the forelegs to the brisket. Peel the pelt back to expose the leg bones. Then, remove them by making cuts at the break joint, which is located just above the foot.

## STUNNING AND STICKING

Several methods can be used to either stun or quickly dispatch a sheep. Inexpensive electrical stunners can be used for only one or two animals. These use an electric current to initiate cardiac arrest to kill the sheep. The animal can then be placed on a table, cradle, or hoist to begin the butchering process.

The simplest method to kill the sheep is to stick the jugular vein with your knife to create blood flow. This can be done by placing the sheep on its side, wrapping the front feet and rear feet together so that the hooves cannot cause you injury, and then placing it on a table or platform with its head draped over the edge.

If you choose to hoist the sheep by its hind legs, tie its front feet together with a rope or cord and then pull it tight toward its hind legs. This will hold the front legs steady, restrain the sheep, and allow you to make a swift, clean kill.

To cut the jugular veins, grasp the jaw or ear with one hand and insert the knife behind the jaw while drawing the blade edge outward and out through the pelt. This will sever the jugular veins and carotid arteries.

One advantage of sticking a sheep versus stunning it is that you achieve a more complete bleed from the body because the heart is still beating. If the sheep is suspended by its hind legs, the flow of blood is downward and will continue while the heart works. Gravity will assist in making as complete a bleed as possible. For lambs, the blood yield may be as much as 3 percent of their live weight.

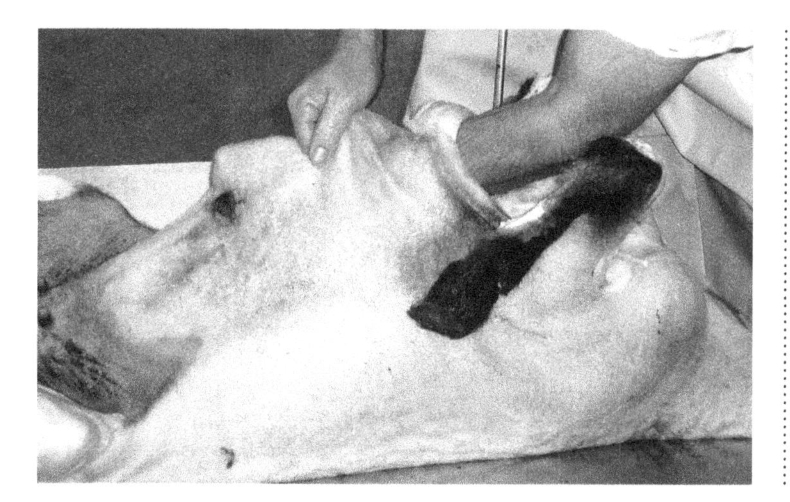

1. Place your fist between the pelt and muscles in the opening at the brisket and push inward to loosen the skin. This is called fisting and is the most effective way to separate the pelt from the body. Continue this motion down both sides while the carcass is still horizontal.

# SKINNING

When the blood has finished flowing from the carcass, you can place the sheep on its back in the cradle or on a table. To begin skinning the carcass, grasp a foreleg and slice the skin open with the point of your knife down toward the chest. Do the same with the other foreleg, having the two cuts meet at the front of the breastbone. Then, finish skinning out both front legs.

Next, skin the hind legs. Begin by holding one leg and open it down the backside from the hoof to the rectum by holding your knife fairly flat as you slice down. Holding it in this way will help avoid cutting the tendon and the colorless connective tissue membrane just under the skin that separates it from the meat in the carcass. All four legs should now be skinless. You can start removing the feet at either the front or rear.

Remove the front feet by cutting through either the break joint or spool joint, depending on the age of the animal. In young lambs, the break joint will be a swelling in the long cannon bone just above the foot. Break joints in yearlings and older sheep are denser and harder to cut. For these, you should remove the foot at the first joint above the hoof.

To remove the hind feet, begin by removing the foot at the joint closest to the hoof. By not cutting it higher, you will leave the backside tendon anchored, which you can use to suspend the carcass. Carefully slice along the leg bone for about 3 inches (7.5 cm), separating tissue holding the tendon to the leg bone. These slits will allow you to insert hooks that will hold the carcass for evisceration. Do the same with the opposite hind leg.

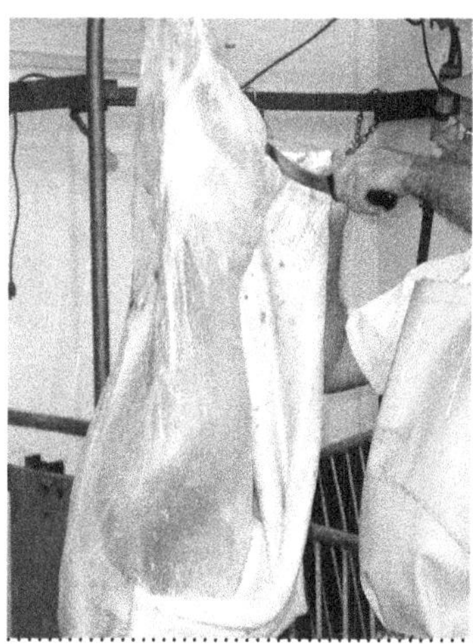

2. Make a circular cut around the anus to loosen the pelt and then pull downward to strip it from the carcass.

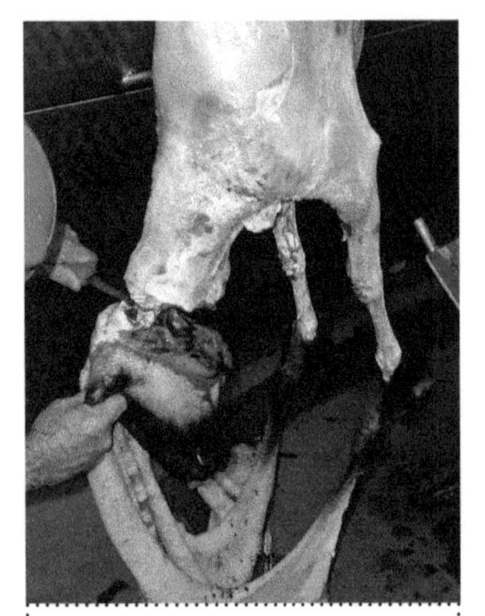

3. The pelt will still be attached to the front feet and head. Next, remove the head by severing it at the atlas joint and finish removing the front feet at the break joints.

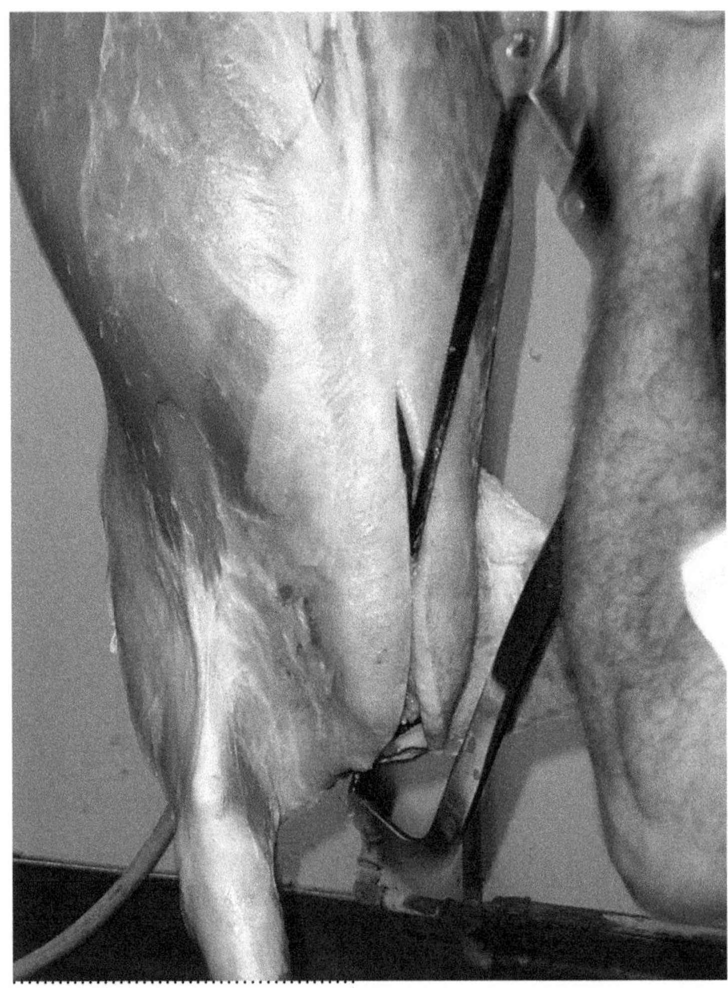

4. To open the thoracic cavity, make a cut through the muscles to the brisket and saw through the sternum bone. Unlike beef, the esophagus and trachea do not need to be separated if the brisket is opened before both are removed.

To separate the skin and fleece from the body, grasp the pelt at the cut, make a fist with your free hand, and slide it forward separating the skin from the body. Push your fist against the pelt and not the carcass as you are loosening it. By repeating this motion, you will loosen the skin without needing to use a knife. This will eliminate cuts and bruises to the body of the carcass. Always have clean hands when loosening the pelt to avoid contaminating the carcass with wool and dirt from the fleece.

With most of the skin now loosened from the body, you can attach hooks to the hind leg tendons and raise the sheep to a level that allows you to comfortably work with the carcass. Once suspended, you can remove the head by cutting behind the jaw and separating it at the base of the skull.

The trachea and esophagus are still attached to the internal organs and must now be separated so that you can remove those organs during evisceration. You will then be able to pull them out when you remove the internal organs. If this separation is not done, you will need to split the brisket prior to evisceration to remove the abdominal and thoracic organs at the same time.

With the carcass suspended, cut open the center of the loosened pelt. Pull the fleece toward you as you slice down the belly being careful not to cut into the abdomen. Loosen all the skin by moving your fist around the entire carcass and up the legs. Avoid pulling or stripping the pelt off the carcass as this may damage the connective tissue membrane by tearing and exposing the muscle.

Sever the anus by cutting across it where it is attached to the pelt. Then, use your fist to loosen it. Finish loosening the pelt at the shoulders. Once the pelt is completely loose, it should easily slide off the carcass.

Rinse the carcass with clean, lukewarm water before opening the body cavity. This will remove any dirt, wool, or other foreign materials that may have attached to it.

This drawing represents the five primal cuts in relation to the skeleton; shoulder, rack, loin, leg, foreshank, and brisket.

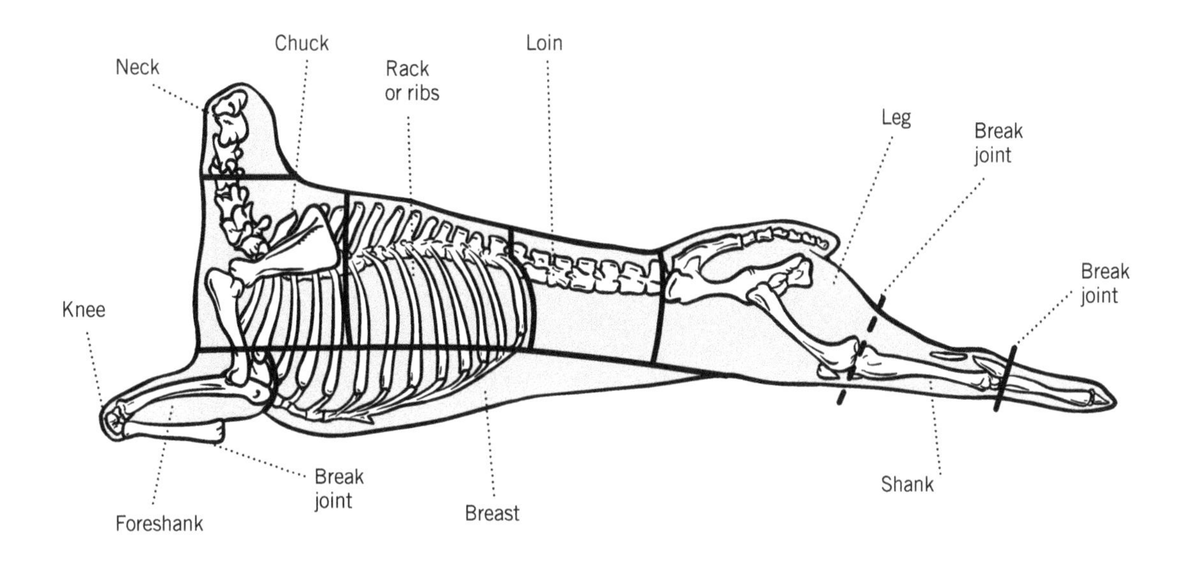

# EVISCERATION

Evisceration of sheep is very much like that of cattle and pigs, except they are smaller. Avoid cutting into the intestinal and digestive tract while opening the body cavity so that it is not contaminated by fecal material.

Place a bucket under the carcass to catch the intestines and blood. Begin by cutting around the anus, loosening it from the pelvis. Cut as close around the pelvic and tail bones as you can until it is free to pull out. You should tie the anus shut with string or a light cord so that any fecal contents do not spill out. Once securely tied, you can let it slide down into the body cavity.

To open the belly, start your knife tip at a point just below the junction where the outer skin of the two hind legs intersects. Pull the skin toward you as you make a cut long enough to insert your first and second fingers to help guide the knife point. Or you can insert the handle into the abdomen cavity the same way as in cattle or pigs to open it.

After the body cavity is open, grasp the tied end of the anus that you let slide into the cavity earlier with your free hand and slowly pull the intestines and organs toward you. Gravity will help pull these from the body, and the bladder and kidneys will also drop as you sever the ureters.

When all the viscera have been removed, split the breastbone with a saw or sturdy knife. Wash both the inside and outside of the carcass with cool water and remove any traces of blood, dirt, tissue, and other foreign matter. You can also trim any scraggly ends or pieces from around the neck or other areas. The carcass is now ready to cool. Chilling the carcass makes it easier to cut up the various parts as the fat within the meat and the muscles become firm.

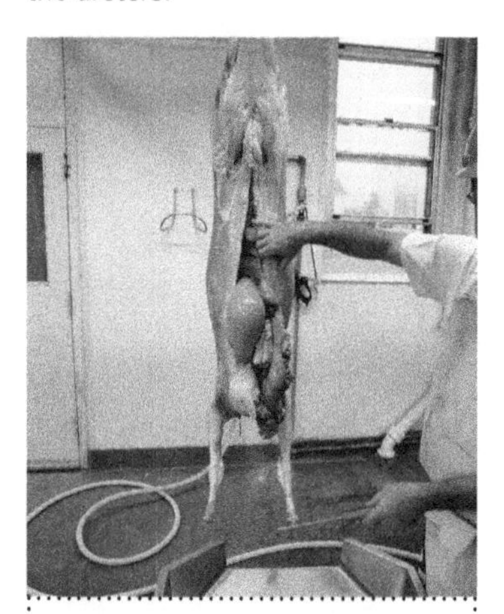

The visceral and thoracic cavities can be cleared in one step and placed in a tub for disposal or inspection. The carcass is now ready for trimming, followed by washing with clean, cold water.

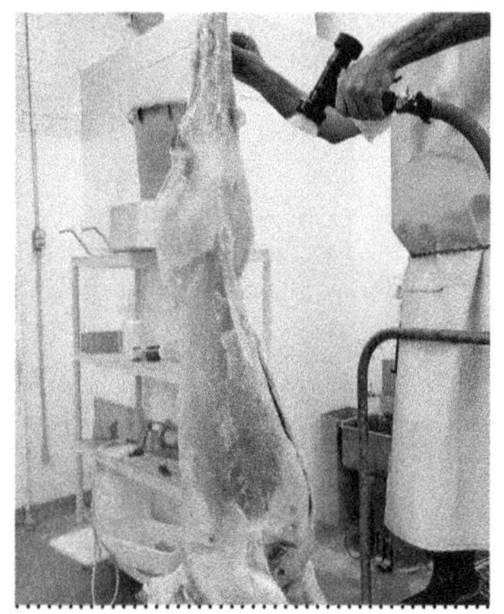

Inspect the carcass and remove any hair, fecal, or foreign matter attached to it. These are the main causes of lamb carcass contamination, which most often occurs during removal of the pelt. Once the carcass is clean, it is ready for washing, cooling, and later fabrication.

# LAMB CUTS

With an average bone-out of about 30 percent meat for a market lamb, you can expect a yield of about 15 pounds (7 kg) from a lamb from a 100-pound (45.5 kg) carcass.

Five primal cuts begin the process of deconstructing the chilled carcass: leg, loin, rack (ribs), shoulder, and foreshank. These can be further broken down into the more valued cuts.

The lamb carcass is not split longitudinally, from rump to shoulder. Rather it is cut laterally, across the body. For home use, this is not necessary because you are cutting the carcass to suit your needs rather than a specific market or other customer.

To split the carcass in half, you will need a meat saw or a heavy, sturdy knife to slice down the backbone. This is most easily done by suspending the carcass by the hind legs. Begin at the aitchbone of the pelvis and saw a straight line down and through the end of the neck.

1. If cooled, lamb carcasses typically are left whole and not split in half before fabrication begins. If splitting the carcass, use a heavy knife or meat saw and slice down the backbone to separate the two sides. If not splitting the carcass lengthwise, you can begin with a lateral cut.

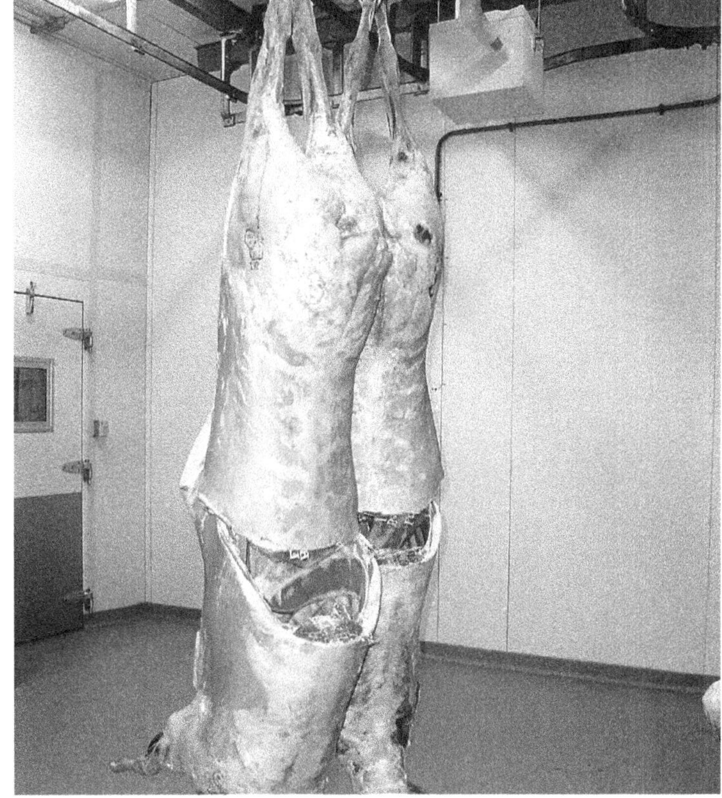

2. Split the carcass by first separating the foresaddle and hindsaddle between the twelfth and thirteenth ribs. The two carcasses shown here have been split but are still hanging in a cooler. Once they are placed on a cutting table, begin by cutting off the legs at the rear hocks and front knees.

If you are going to process the carcass during the same session, you can leave one half suspended while you cut up the other. Lay one half on a clean, sanitized table, making sure that the equipment you use has been thoroughly cleaned and sanitized. The process of cutting up a lamb carcass allows you to break down large sections into smaller pieces for further cutting. Try to cut it up in a room with a cool temperature to keep the meat from becoming too warm.

Begin by dividing the half into two parts: the foresaddle and the hindsaddle. This roughly cuts it in half. The foresaddle consists of the shoulder, rack, foreshank, and breast and makes up about 51 percent of the carcass. The hindsaddle composes the loin, leg, and flank and represents about 49 percent. To separate them, make a lateral cut across the carcass between the twelfth and thirteenth ribs.

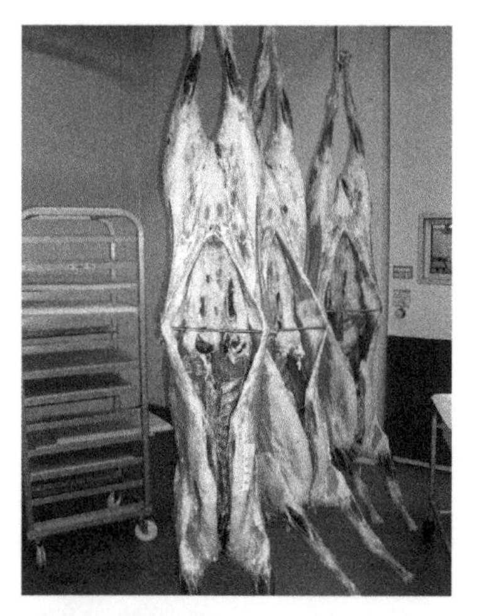

3. There are thirteen ribs on a lamb carcass. To separate the rack from the shoulder, count eight ribs from the posterior end and cut through the muscle with a knife; finish cutting through the backbone with a saw. There will be four ribs remaining in the shoulder.

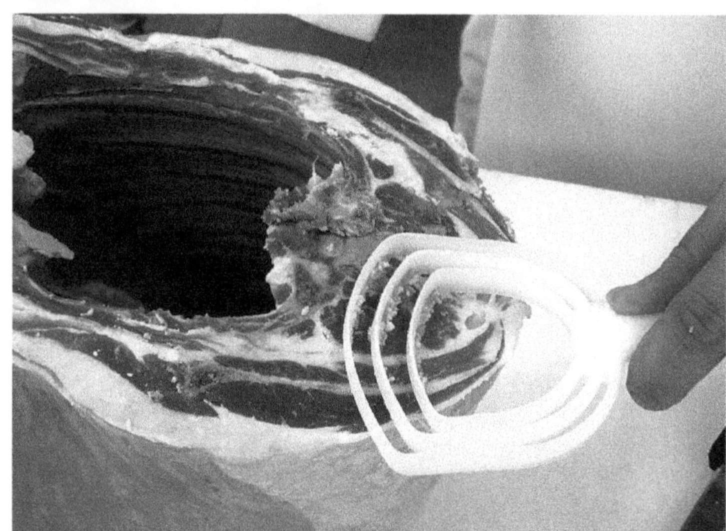

4. One very useful tool is a plastic bone scraper. This helps scrape away bone dust caused by the blade sawing through bone. Bone is a good medium for microorganisms to grow and reduces the shelf life of meat. Scraping away this bone dust will minimize those effects.

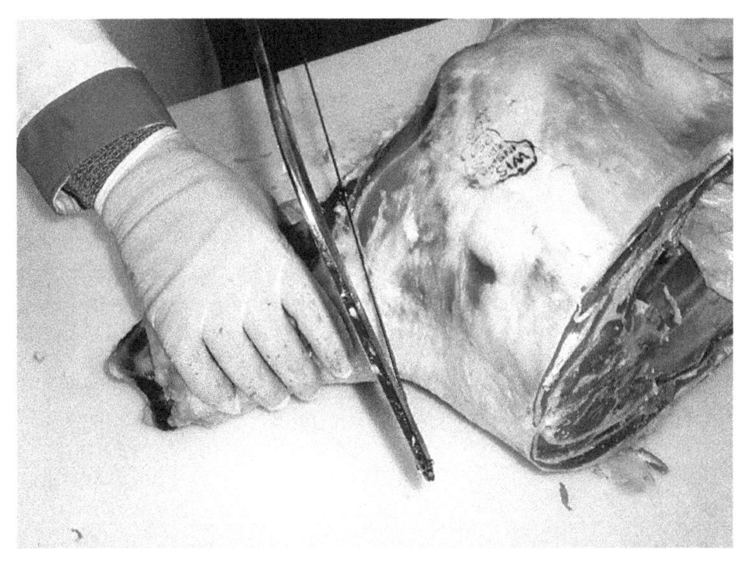

5. Remove the neck with a lateral cut just in front of the shoulder. It can then be cut into neck slices for roasting or braising, or trimmed and used as stew meat, ground lamb, or for sausage.

## LIVE ANIMAL LAMB CUTS

- 12 percent from legs
- 4 percent from loin
- 4 percent from rack
- 10 percent from shoulder
- 30 percent total meat

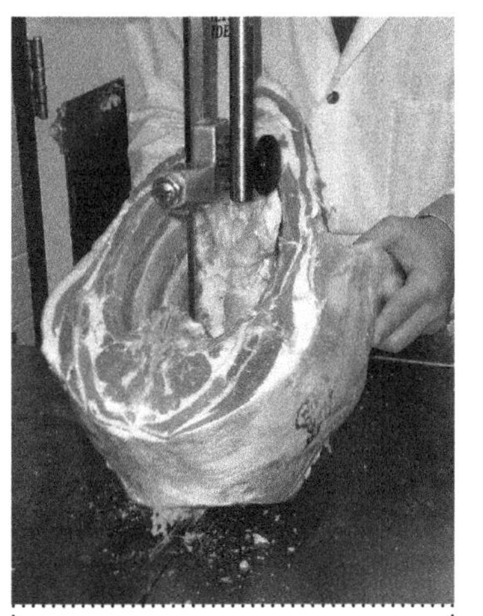

6. Split the shoulder in half and trim away the fat. Cut off the breast where cartilage appears and then remove the remaining portion of the foreshank. This should expose the brisket bone, which can be pulled out by hand.

## JUDAS GOAT

Sheep have a natural flocking instinct and will huddle together or follow the leader of the group, particularly if they perceive an outside threat. Because of this behavior, they can be manipulated in their movements by using a goat. At some packing houses, goats are usually reliable and will return time after time to lead lambs to slaughter. Any goat that fills this role is known as a Judas Goat for its deception and betrayal of the lambs.

## FORESADDLE

Slightly larger than the rear of the carcass, the foresaddle contains four major portions that can be further reduced. Begin by dividing it into two pieces that contain the shoulder and rack (ribs). This cut is made between the fourth and fifth ribs and will leave you with an eight-rib rack for later. At this point, the breast and foreshank are still attached. They should be separated from the shoulder and rack. Remove them by sawing across the arm bone at a point slightly above the junction of the arm bone and the foreshank bone.

The shoulder is the largest cut in the foresaddle and contains a number of bones that make it more difficult to carve and slice. However, if properly cooked, shoulders provide a delightful meal.

Remove the neck from the shoulder by cutting across it laterally, leaving about 1 inch (2.5 cm) of neck on the shoulder. The neck then can be sliced into small pieces.

The shoulder is often called a square-cut because it fits the dimensions of a square. It includes the blade face, which has a surface mostly of bone and lacks muscle, with part of the blade cartilage. Any chops removed from the shoulder are called blade chops because of the presence of the scapula, or shoulder blade. The most forward portion of the ribeye will extend through this area.

The arm chops come from the arm face that is positioned at a right angle to the blade face. This is the muscular part of the shoulder, but it is less tender because these muscles do a lot of work in providing locomotion for the live animal. Typically, muscles that have more use are less palatable and also contain a considerable amount of connective tissue to other less-used muscles.

Instead of making chops, you can cut out the rib cage and then the ribeye by following the natural seam. This is the boneless blade roll that can be held together with skewers. You can also cut out the shoulder bone and use it as a boneless roast. The rest of the shoulder pieces can be diced for stew meat or kabobs after the bones have been removed.

## HINDSADDLE

The hindsaddle contains the most valuable cuts of the sheep or lamb. These include the loin, leg, and flank. Start your cuts by removing as much of the flank as possible. Do this by making a cut about 2 inches (5 cm) away from the loin eye. The flank is a large, flat, straight muscle that, in beef, is cut into flank steaks. Because of its size, texture, and flavor, it is best used for ground meat with sheep and lambs, or it can be rolled to be roasted.

The leg is the largest cut and represents about one-third of the lamb carcass. When you cut the leg from the loin, the sirloin will be included with it. Although typically referred to as a leg of lamb, this term also implies that it is the whole leg with the sirloin still intact.

To separate this primal cut from the loin, sever it at the seventh, or last, vertebra. Use a saw to make your cut, as this will result in a flat face surface on the loin. To remove the tailbone with your knife, make the cut so that it leaves three tail vertebrae on the leg.

You can trim off any outside fat, but avoid removing the thin membrane that separates the pelt from the muscle. This membrane holds the shape of the leg and helps retain moisture and juices during cooking.

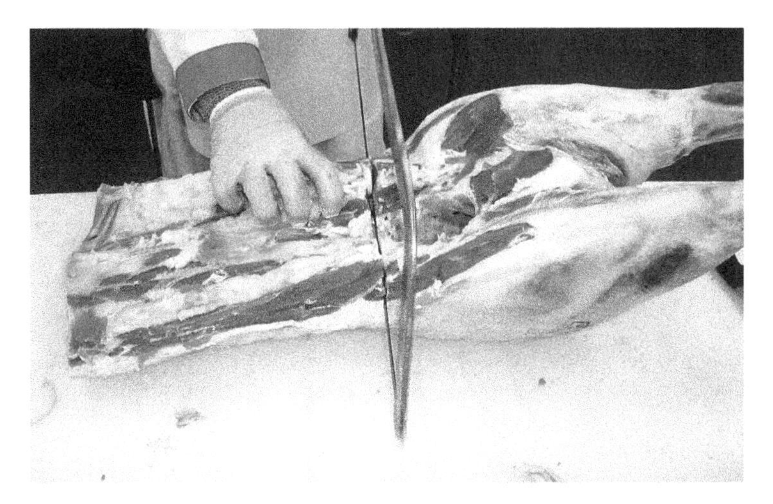

1. Before cutting the hindsaddle, remove as much of the flank as possible by making a cut 2 inches (5 cm) from the edge of the loin through the thirteenth rib and following the cut parallel to the loin and along the leg. These can be cut later into flank steaks or ground meat. Remove the leg by making a cut through the last lumbar vertebrae. Use your knife for your initial cut and finish with your saw.

To remove the leg, cut the Achilles tendon, the large tendon above the hock where it attaches at the base of the leg muscle, leaving the other end attached to the hock. Cut through either the hock joint to remove the lower part of the leg or through the break joint that is located about 1½ inches (4 cm) above the hock. In lambs, you may be able to use your knife to break this soft area, but it will generally require the use of a saw for older sheep or goats.

The hindshank is composed of a large amount of connective tissue, which can be identified as white, silvery streaks. Because this is a less tender portion, it is best used as ground meat. You can remove the shank muscle by cutting through at the stifle joint, which is the second joint above the foot. Once separated, you can trim the meat from the bone and set it aside for grinding, or it can be cut into smaller pieces for braising. The shank can also be roasted.

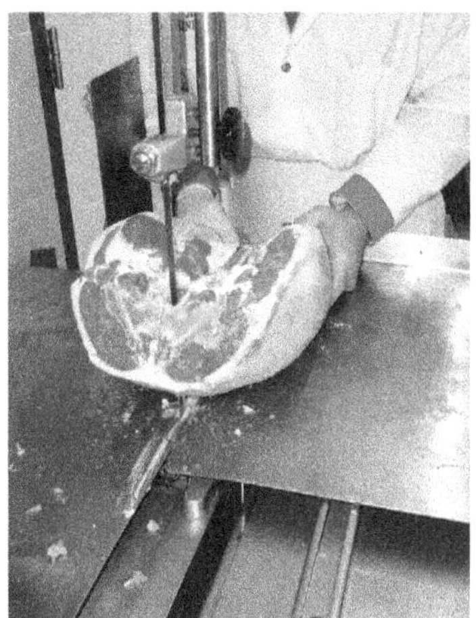

2. Split the leg portion of the hindquarters into halves by sawing through the aitchbone.

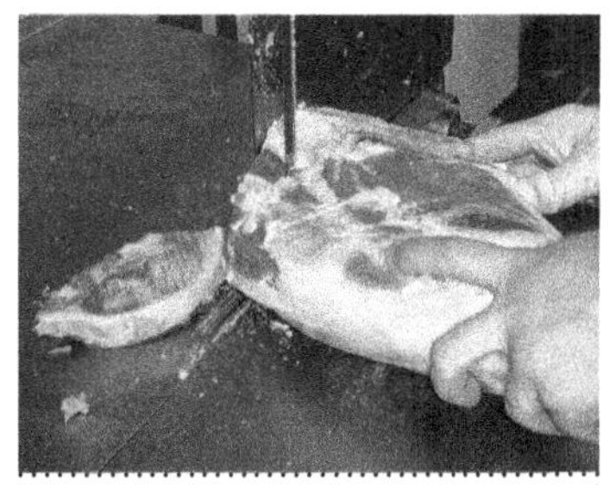

3. You can remove the leg (shank portion) first if using a band saw. Otherwise, it can remain and be used to hold the lamb leg as you remove leg sirloin chops. There are two to three sirloin chops per leg per side. They are recognizable because part of the pelvic bone will be in them.

The sirloin may be removed from the leg with lateral cuts made into sirloin steaks. The largest of these will be located at the top of the sirloin. You can saw the whole leg into chops and steaks. This will result in four to six sirloin chops, depending on how thick you wish to make them. These chops are similar to beef sirloin steaks in bone and muscle structure.

Because of the pelvic bone, the rump is not sliced or made into steaks. However, by trimming and removing this bone, the meat can be used for kabobs, stews, or ground into patties.

The lamb's legs may be separated into sirloin and shank halves, and these may be further broken down if desired. The leg can be either boned out or made into steaks. The boneless leg then can be roasted or tied into a rolled roast.

To remove the pelvic bone, cut around the ball where it joins the leg. You will be able to trim it out and then separate it from the leg bone at the ball joint. Remove the shank at the stifle joint, which will allow you to cut around the end of the leg bone until it is loosened.

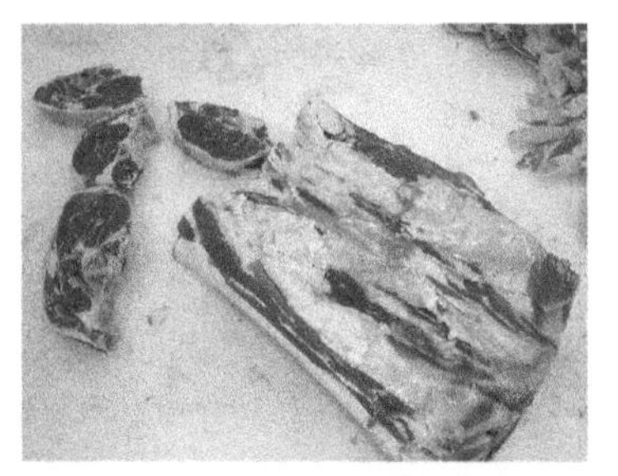

1. The loin is the most valuable cut of the lamb carcass and will vary in size from end to end. It can be cut with the backbone attached for loin chops or completely cut out to make a boneless loin roast.

2. Split the loin down the spine into two separate halves being careful not to cut into the loin eye. You can make 2-inch (5 cm) lateral cuts to create chops. The bones can remain intact or removed as desired. Removing them creates boneless chops or a boneless loin roast.

3. The ribs can be cut into four sections from the two carcass sides. Each of these can be grilled or roasted as a rack of ribs or riblets.

You should be able to remove the entire leg by pulling on it without cutting any of the muscles. There is a lymph node, usually surrounded by fat, located at the rear end of the boned leg between the bottom and the eye muscles. Carefully trim this node off without cutting into it. With the leg now finished, you can begin work on the loin.

## LOIN

The loin is the most valuable cut in the carcass because it contains the most tender muscles. The loin and rack (ribs) often compete for the higher price because there are so few of them from each animal. Only 4 percent of the live animal will end up being cut into loin chops.

From end to end, the loin will vary in size and shape. At the seventh lumbar vertebrae, the loin eye is oval-shaped and the tenderloin is at its maximum size. At the front end, the loin is larger and more symmetrically shaped.

You can cut the loin into chop widths of between 2 to 3 inches (5 to 7.5 cm), leaving the bone, which looks like a T, and is similar to the T-bone steak in beef cuts. Or you can remove the vertebrae to produce a boneless loin chop, which can be rolled, netted, or tied. You can also leave the loin intact for a loin roast.

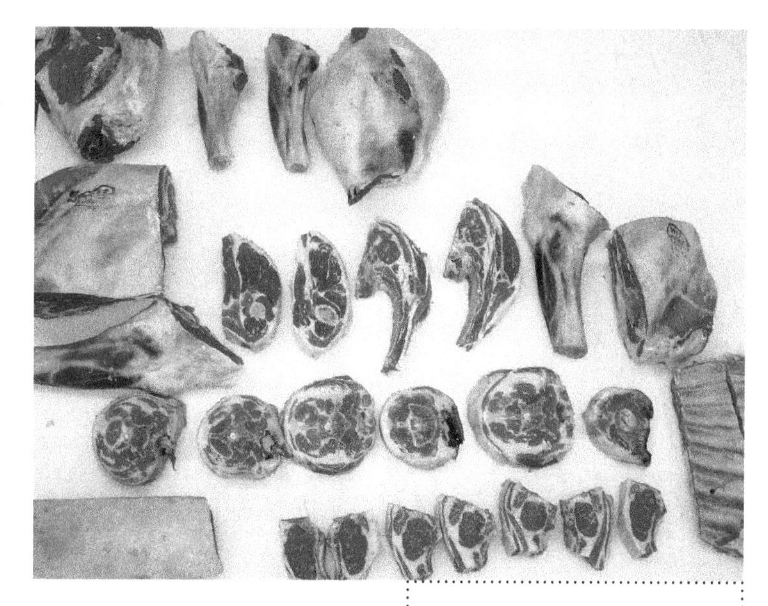

Lamb carcasses can create a wide variety of choice cuts, including roasts, chops, loin and rib chops, ribs, racks, and legs. Lamb cuts are increasing in popularity and provide an interesting alternative to beef and pork.

## ROUGH CUTS

There are several pieces still left to cut, including the flank of the hindsaddle and the foreshank and breast of the foresaddle. These account for about 10 percent of the carcass weight but typically have more fat than muscle. Trim as much of the fat away as possible and use the meat pieces for grinding into patties.

The breast contains rib bones and the breastbone. You can cut each rib apart for riblets or you can cut them into sections of several ribs together for lamb spareribs. There is a thick muscle in the breast, which corresponds to the beef brisket, that can be trimmed and used for cubes.

The foreshank can be braised or trimmed and cubed for stew meat, or ground into patties. When finished with the first half of the carcass, you can begin with the opposite half to make the same cuts.

CHAPTER 5

# PORK

**THREE ERAS STAND OUT FROM A MODERN HISTORICAL PERSPECTIVE IN PORK PRODUCTION THAT EXPLAIN AS MUCH ABOUT SOCIETY'S TASTE PATTERNS AS THEY DO ABOUT FARMING TRENDS. THESE CAN BE DESCRIBED AS THE ERAS OF LARD, MEAT AND BACON, AND LEAN WHITE MEAT.**

A healthy, well-grown, 200-pound (91 kg) pig is a good choice for home butchering because it will yield about 100 to 110 pounds (45.5 to 50 kg) of eating meat.

You can raise the pigs yourself, purchase them live from a grower, or purchase a dressed carcass from a local butcher and fabricate the carcass yourself.

Lard once found many uses, such as in making candles, soaps, and cooking fats. As lard demand decreased with the advent of petroleum products, vegetable oils, electricity, and consumer dietary changes, the trend moved away from fat hogs to leaner ones with a higher ratio of muscle to fat. A period followed where the fat composition in pork meat dropped so dramatically that consumers found it lacking in flavor and moved more toward chicken consumption for their white meat. Today, a more tasteful pork product is produced because of a favorable balance between fat and lean.

A 200-pound (91 kg) pig with a typical 72 percent dressing weight will yield a carcass of about 145 pounds (66 kg), or about 73 pounds (33 kg) per each half or side. This will include meat, bones, and fat. The cutting yield is the amount of meat you get from the total carcass. Using a typical cutout rate of 60 percent for pigs, this 145-pound (66 kg) carcass will yield about 110 pounds (50 kg) of meat for your use. The other 35 pounds (16 kg) will include fat trim, bones, and skin.

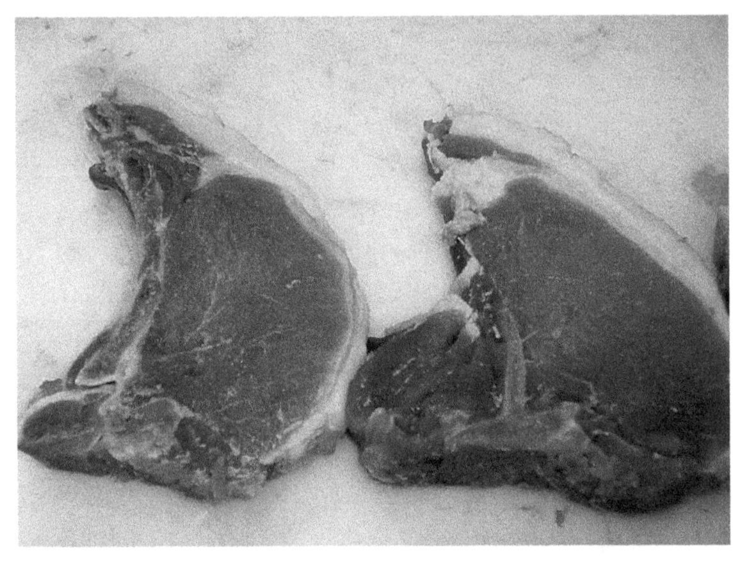

1. Loin chops are cut to ½ to 1 inch (1.5 to 2.5 cm) thicknesses. The loin chop has a portion of the loin on one side and tenderloin on the other, and it is located toward the junction of the leg and loin. The rib chop (on left) only contains the loin muscle.

The largest part of the carcass is usually the ham, which can be about 23 percent of the live carcass, but 18 percent of a dressed one, or in this case about 27 pounds (12 kg). The side or belly and the loin areas represent about 15 percent each, or about 40 pounds (18 kg). The picnic and Boston butt are each about 10 percent or 16 to 20 pounds (7 to 9 kg), and the miscellaneous portions—including the jowl, feet, neck bones, skin, fat, bone, and shrink—account for about 25 percent of the carcass weight. In our example, this would amount to about 38 pounds (17 kg), which is well over a third of the entire carcass. There may be some variance among pigs; these percentages generally hold true for normal, well-developed pigs of that weight range.

## PORK CUTS

The five major areas where cuts are derived can be further broken down into cuts often found in retail markets: the picnic shoulder, Boston butt, loin, ham, and belly or sides. One reason that retailers or you, if you are marketing your meat to others, decide to charge higher prices for certain cuts is because of supply, demand, taste, and ease of cooking. For example, pork chops typically are in demand and relatively easy to prepare. They sell quickly while other cuts may not move as fast.

The picnic shoulder includes the upper front leg above the knee. This cut lies just below the Boston butt and contains a higher level of fat than the other cuts, which makes it a flavorful and tender portion. The picnic shoulder can be smoked and cured to make the picnic ham, which is then ready to eat cold or hot. The arm and shank bones make up the shoulder and create a high ratio of bone to lean meat. When well-trimmed, this cut is used for lean ground pork and can be cubed or cut into strips to use for kabobs, stir-fry, or stews.

The Boston butt, also called the shoulder butt, is often a better cut than the picnic shoulder. It lies at the upper portion of the shoulder from the top to the plate to make the backbone. This cut is tender, full of flavor, and can be cut into roasts with the bone intact or cut out for boneless roasts. The roasts can be cut into blade steaks that can be broiled, grilled, or braised.

The pork loin cuts are located directly behind the Boston butt and include a portion of the shoulder blade bone. The loin includes most of the ribs and backbone all the way to the hipbone at the rear. There is a loin area on each side of the pig and together will account for about 20 percent of the carcass weight. Many retail cuts are derived from the loin, including top loin roasts, pork chops, baby back ribs, pork tenderloins, loin chops, rib chops, and blade chops.

The loin constitutes a long strip that contains the top section of the ribs. When these are trimmed away, the result is a boneless pork loin. The section of the loin between the blade end and sirloin end is usually referred to as the center, as in center chops and center roasts. A boneless pork loin is smoked to produce Canadian bacon. Rib bones trimmed from the loin can be barbecued as pork back ribs. This is also the area where pork backfat is located. This is the thick layer of fat between the skin and the eye muscle, which may have some cooking uses but is used mostly to help determine carcass grades.

3. The meat and fat trimmings from pork can be used in several different ways, including mixing in sausage, particularly with wild game, to add texture and flavor.

2. The sirloin chop can be identified by the round bone or flat bone shape. The rib chop contains part of the rib bone the sirloin does not.

Hams make up about one-quarter of the carcass weight and come from the rear leg area of the pig and include the aitch, leg, and hindshank bones. This is a prime cut area of the pig because it contains little connective tissue, making it more flavorful whether it is cooked, cured, or smoked. Hams can be deboned, and the shank portion of the ham, called the ham hock, is used the same as the shoulder hock.

The pork belly or sides are where the bacon and spareribs are cut. They are located below the loin on each side and account for about 15 to 20 percent of the carcass weight. This area contains a lot of fat with streaks of lean meat. It also provides the spareribs, which are separated from the rest of the belly before cooking.

The miscellaneous portion of the carcass includes the jowl, pig's feet, tail, neck bones, skin, and fat. Some cooks highly prize these areas, but the parts are often dismissed by the general public as unusable.

Eating raw pork is strongly discouraged because of the presence of a parasite *Trichinella spiralis* that causes trichinosis in humans. These little roundworms migrate into pig muscles and mature into an infective stage. Unless destroyed by minimum cooking temperatures of 140°F (60°C), they still will be viable parasites for infections. Most experts recommend a minimum of 150°F (66°C) for cooking because of the inaccuracy of many thermometers. Trichinosis infections in humans can cause nausea, diarrhea, muscle pains, and aching joints. It is a treatable infection but nevertheless is an uncomfortable experience.

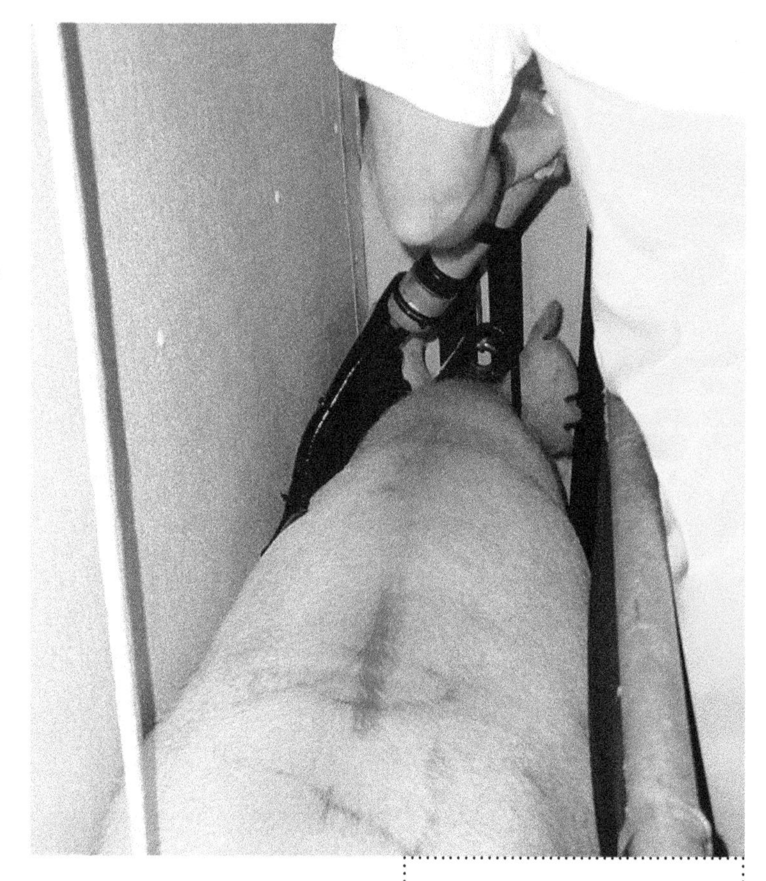

4. Begin with a clean work area before bringing the pig in to kill. The area used should be as free of dirt, dust, flies, and insects as possible. When everything is ready, you can use an electric stunner or a gun to immobilize the pig and to render it unconscious.

## BUTCHERING AT HOME

Whether you are butchering at home or having a pig processed at a local meat slaughter plant, you will have to decide which pig to use if you have more than one to consider. Pork carcasses for home meat use should be the highest quality pig you produce and one that is from five to eight months of age. Pigs fed liberal amounts and quality feeds grow rapidly and will produce pork of proper size and finish.

Butchering at home requires some thought before, during, and after the entire process. Planning ahead for this event will minimize mistakes, reduce the chance of injury to helpers, and provide you with quality meat products for your family.

Develop a list of all the steps needed, from beginning to end, several days before butchering your pig. Planning ahead will reduce any surprises during the butchering process and help organize the event in a logical, efficient manner. Having a list of the equipment and butchering tools required will ensure everything is at hand when needed. Attention must be paid to cleanliness at all times so the meat isn't contaminated during any part of the process. A review of the skeletal structure, digestive system, the placement of organs, and some understanding of the circulatory system will be useful when you are cutting the carcass. Knowing the precise location of the jugular veins with in the neck will help you make a clean, swift kill.

The location where you plan to carry out the kill should be properly equipped for the job. A shed or building that is free from dust or outside elements can provide a good place for the initial stages. If this is used, construct a small holding pen near the butchering table to reduce the distance the carcass has to be carried. If you decide to butcher the pig outside, a sheltered pen can be built near the area you will work to cut it up.

There are two ways to remove the skin and hair. One is by manually skinning the pig, which actually requires less effort, and the other is by using a scalding procedure. These will be detailed later in this chapter, but keep in mind that if you choose to use the scalding method, you will need a convenient heating arrangement, such as a scalding vat, and an efficient way of swinging the carcass into the boiling water with a block and tackle or some other apparatus.

A proper set of butchering tools includes sticking knives, skinning knives, boning knives, butcher knives, a steel sharpener, meat saws, and meat hooks. Other useful items include thermometers, a meat grinder, meat needles for sewing rolled cuts, hair scrapers, hand wash tubs, clean dry towels, soap, and vats with hot and cold water.

Providing a proper location and sharp tools will aid in more efficient slaughtering and less time spent looking for items at critical moments.

## CARE BEFORE BUTCHERING

Two to three days before butchering, confine the pig in a small solitary pen. Provide plenty of fresh water, but restrict feed 24 hours before butchering so the pig has less material in its stomach and intestines. Providing a cool and calm environment several days beforehand will keep the pig rested and quiet. Never attempt to butcher a pig that is overheated, excited, or fatigued. When the body temperature is above normal, the meat easily becomes feverish and is difficult to chill properly; poorly chilled meat cannot be properly cured. This increase in temperature can cause the meat to spoil or be tainted before it is cut up.

Some spoilage and low-quality meat can be directly attributed to natural forms of bacteria that have been allowed to develop and multiply. The bacteria that is found in the blood and tissues of a live pig must be held in check to prevent it from multiplying until the meat is cured. This is one reason butchering was historically done in the early spring or late fall of the year when the weather was cool. Think of it as a race between the bacterial action in the blood and tissues that want to multiply and the curing agents used, such as salt, cold water, and other factors, which depress bacterial growth. You need to win this race.

## DISPATCHING YOUR PIG

When the butchering tools are laid out, the table is thoroughly washed with soap and dried, enough help is on hand, dripping pans are ready to catch blood, ice is in the cooling vat, and everything is in place, you can dispatch your pig.

1. Stun or shoot the hog at a point near the intersection of the two imaginary lines, just above the eyes and at the center of the forehead.

This is perhaps the most critical time in the whole process, and if you feel uncomfortable sticking a knife into the throat of a live pig, you may want to have someone else adept at the task handle this part of the process. This should be arranged prior to this day and not be a last-minute decision.

Butchering a pig by only sticking it with a knife is the most practical, efficient, and humane method of killing a pig. Other methods, such as shooting, are less reliable unless the pig is very agitated. A good bleed is difficult to obtain when the pig is shot because the heart stops and no longer pumps blood through the body. If you shoot the pig, you will need to rely on gravity to drain most of the blood from the carcass rather than being assisted by heart action.

You can stick the pig either in a raised or prone position, depending on whether or not you have equipment to raise it in the air. To hoist the pig in the air, a chain or straps can be looped between the hock and the hoof in order not to bruise the hams. The pig has less ability to free itself if hanging upside down than if it is rolled on its back and the feet are held by several people while one person sticks it. This upside-down posture tends to immobilize the pig and makes the actions with your knife easier and safer for you. The most satisfactory bleed occurs when a pig's head hangs downward. Be aware that this position will be very uncomfortable for the pig and it will typically flail its feet that are free. The feet must be firmly

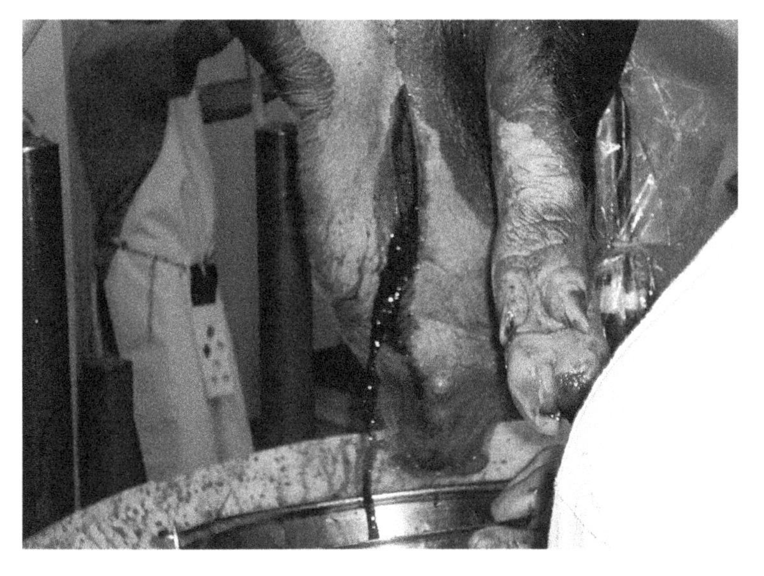

2. After the pig is unconscious, tie the rear feet with a chain or sturdy rope and raise it. Use your sticking knife positioned between the lower jaw and breastbone and press the blade deep into the center of the neck. Make a small vertical incision to sever the jugular vein. This will release a large quantity of blood, which can then be caught in a pan or tub.

3. After the bleeding has stopped, place the carcass on a table or rack. The V-shaped trough shown here can be made for use with pigs, sheep, deer, goats, or veal calves.

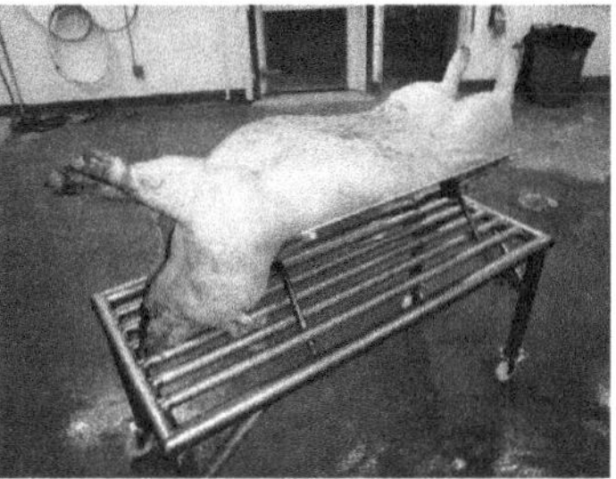

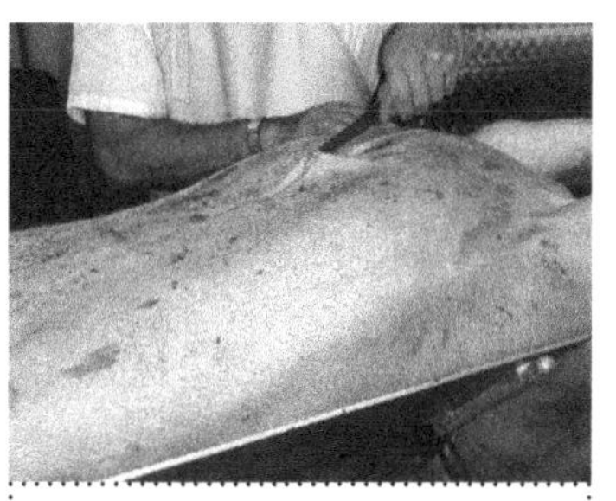

4. Skinning is the easiest way to remove the hair and hide if it is not used for tanning. Begin by making a vertical cut down the midline, knife blade out, of the carcass from the jowl to the pelvis, but avoid cutting through the abdominal wall.

immobilized, as they can be used by the pig to defend itself.

When the pig is safely immobilized, press the sharp blade edge of the sticking knife in front of the point of the breastbone and quickly slide it in to make a short vertical incision about 4 inches (10 cm) long in the center of the neck. This should sever the jugular veins and release large quantities of blood. Having tubs placed below to catch the blood will make cleaning up easier and allow you to work around the pig in dry conditions. The knife should not be inserted too far into the neck so that it enters the chest cavity, as this will cause internal bleeding and blood clots.

Do not stick the heart, as it is needed to continue working properly to pump out the blood as rapidly as possible. Cutting the heart will cause internal bleeding and create lower-quality meat. The key to this phase is to get a good bleed as quickly as possible. When the blood flow has stopped or slowed to a drip, your pig can be moved to a table where it can be skinned, which requires less time and effort than scalding it.

1. Remove the front feet by cutting at a point just below the back of the knee joint. Severing the tendons will allow you to break the joint forward. Then, cut completely through the exposed joint to sever the foot. Do the same with the rear feet.

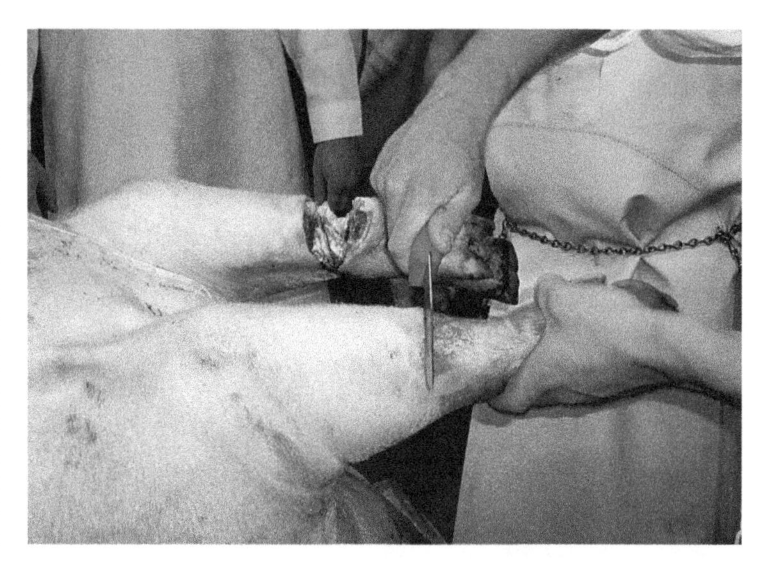

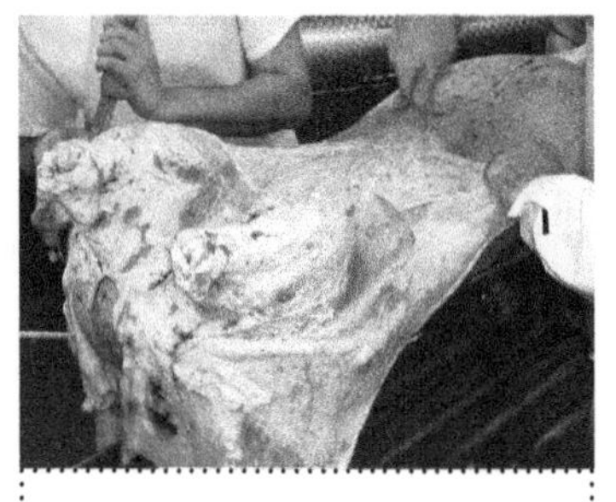

2. Begin the skinning process by pulling the skin up and away from the carcass, and make slow, sweeping motions between the skin and body with your skinning knife. Applying an outward pressure with your knife blade while skinning will help avoid cutting into the carcass.

## SKINNING

Skinning is the removal of the hide from the pig and takes off the outside layer without using hot water or the extra effort of scraping the hair. Most home butchering will not require the use of the skin, which is generally discarded. In the past, the skin was left on the bacon and hams to protect them, but this is not required with modern refrigeration.

When skinning a pig, it's important to remove the skin from the belly without puncturing the abdomen with your knife. This is best accomplished by laying the carcass on a table or trolley for suspending it at an appropriate work height.

With a short skinning knife, begin your first cuts at the rear ankles and slice completely around them, but avoid cutting the tendons above the hocks. Cut down the inside center of each leg to a point below the pelvis and avoid deep cuts into the meat. Do the same with the front legs and cut to a center point at the base of the chest. Use your knife to score a line down the center of the belly, from the anus to the base of the chest, without penetrating the abdominal wall. Start at the chest and create tension on the skin by pulling it away with one hand as you slice with the other. This tension will help separate the skin from the body.

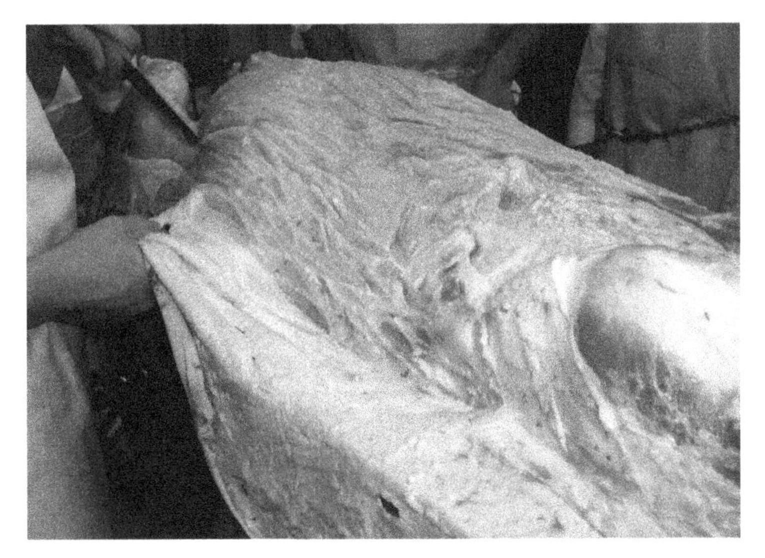

3. As you work the length of the carcass, avoid contaminating it with hair or dirt from the skin. Have clean water nearby to wash your free hand. Also, wash your knife frequently. If the carcass is on its back, you will need to turn it or raise it to finish removing the skin. Be sure the surface where you lay the partially skinned carcass is clean.

When finished with both front legs, start on the belly and slowly work to the rear, pulling the skin away from the center until you reach the base. Do the same with the other side. Start at the top of the rear legs by pulling the skin over the hams. At this point, the skin should be loose from the belly and legs, and by pulling downward and slicing the skin, the weight of the skin will create tension to help with the rest of the process. Once the skin is completely removed, it can be set aside.

## SCALDING

If you do not want to use the skin for any further processes, such as tanning, *you can dispense with scalding the pig*. You can simply trim the skin off with a knife. If you choose to scald the carcass, having the proper equipment makes the job easier, and you can use a tank or barrel for this step. The tank needs to be filled with water brought to a boil prior to immersing your pig. The water can be brought to a boil by using a pit fire underneath the tank, a gas or propane heater, or other means to safely raise the water temperature. This saves time in keeping the process moving because it is difficult to raise the water temperature once the pig is immersed in the water. The water should be kept at between 150°F to 160°F (66°C to 71°C).

If you are using a long horizontal tank, rotate the carcass until the hair starts to slip. If using a barrel, first lower the head into the water while the feet and legs are dry. Then, turn it around, placing the meat hooks in the lower jaw and lower the rear end into the boiling water. This is a more difficult method but will work if no other tank is available. Using an accurate thermometer will help you maintain the temperature, making the scalding easier and eliminating the chance of the hair setting tight against the skin. After lifting the carcass from the scalding water, wash it clean with hot water, scrape off any remaining scruff, and rinse it down again with cold water.

Be very careful when working with boiling water, as any spills can be harmful to you or anyone else helping. An accidental tipping of a barrel of boiling water can create hazardous conditions.

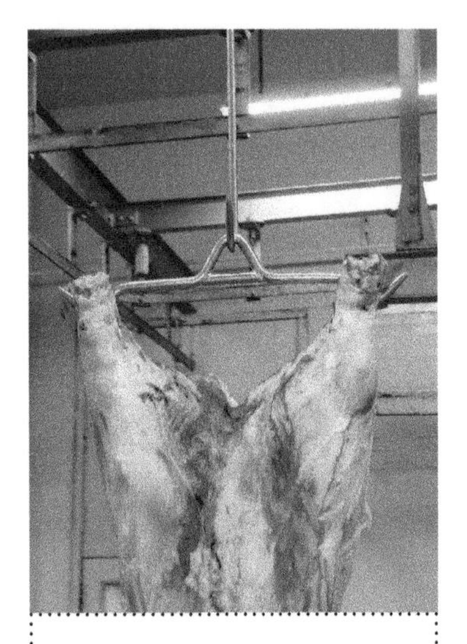

Use a clean chain, rope, or gambrel to lift the carcass to allow easier removal of the remaining skin. To use a gambrel, make a slit in front of the rear leg bones without cutting the tendons. Place the gambrel points in the slits and raise the carcass. The tendons are strong enough to hold the heavy carcass while suspended if they haven't been cut.

The head can be removed before skinning or after the carcass is raised. If left until skinning is finished, remove it by making a cut behind the ears between the axis and atlas joint and around the lower jaw to sever it. The axis joint is the first cervical vertebrae. Then, remove the cheek muscles, tongue, and fat.

## SCRAPING

Scraping the hair off the skin is the next step of the butchering process, and some of it can be done while the pig is still in the scalding vat when you are ready to lift the carcass out and place it out on the table. Using a scraper, start first at the head and feet, as these areas are the first to cool. Your scraping strokes should go in the direction the hair lays, as it will come off easier. After the hair has been removed, use your scraper in a circular motion to work out dirt or scruff that may be imbedded in the skin. A soft bristle brush is useful for cleaning up the carcass once the hair is removed. Any stray bristles of hair can be removed with a little hot water and a sharp knife.

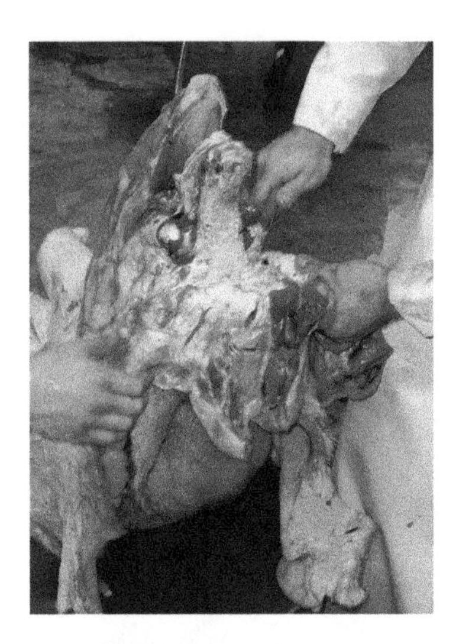

## HANGING

If the pig is laid out on the table, locate the area between the foot and the hock on the rear legs. Make deep cuts up the center of the bone on each leg to find three tendons. Use your fingers to pull the tendons out and slip the gambrel stick through one tendon and then the other. These tendons are strong enough to hold a hanging carcass while you open up the body cavity. Before you make any cuts and incisions to open the carcass, be sure all knives and butchering tools have been scalded and cleaned. Any knife or other tool to be used should be scalded again before use if they have been dropped on the floor. From here on, cleanliness is absolutely essential.

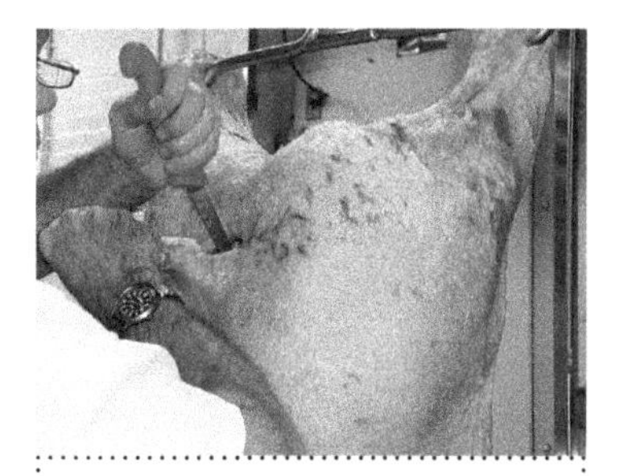

Begin evisceration by cutting around the anus to loosen the muscles holding it. Tie the anus shut.

## REMOVING THE HEAD

Removing the head first accomplishes two things: It gets it out of the way and aids in quickly cooling the carcass. It also permits blood to completely drain from the carcass. Begin by cutting above the ears at the first joint of the backbone and then across the back of the neck. When you reach the windpipe and throat, cut through them and the head will drop, but don't slice the head completely off just yet. Pull down on the ears and continue your cut around the ears to the eyes and then toward the point of the jawbone. When you slice through the last part of the skin at the end of the jaw, the head will come free, but the jowls will still be attached. Wash the head quickly and trim it as soon as possible.

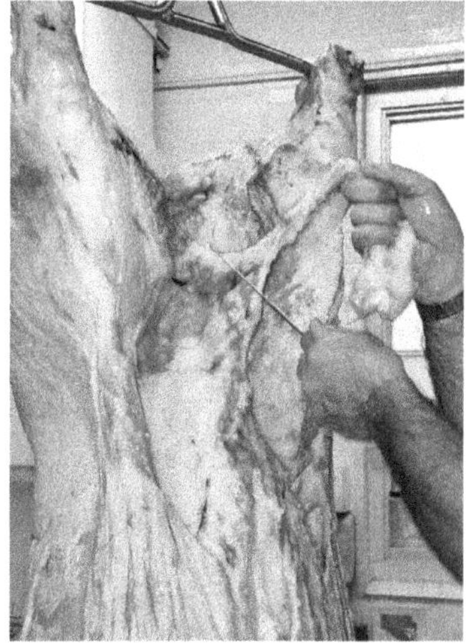

## SPLITTING THE CARCASS

Splitting the carcass is easiest to accomplish when it is suspended. Cut a clean line down the center of the belly between the hams to the sticking point at the base of the chest, but do not cut through the belly wall. To split the breastbone, place the heel of your knife against the bone and cut outward. You may have to work the blade to split the breastbone and divide the first pair of ribs. If your knife will not cut through the breastbone, you may need to use a saw to cut it.

In either case, you should avoid cutting past the upper portion of the breastbone and into the stomach. This is a thin area, and you do not want to cut the stomach open. By cutting through the breastbone and first rib, you will open the chest cavity sufficiently to allow any blood that has accumulated to drain out.

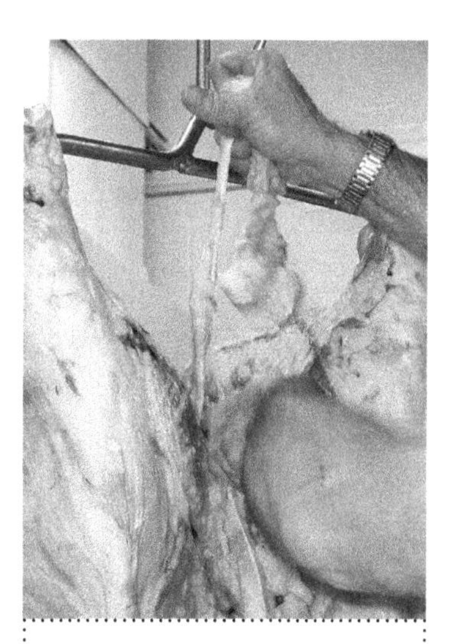

1. A castrated male pig, or barrow, will have the penis and sheath still intact. This should be removed like the skin by cutting upward toward the anus until it is severed. It can be finally removed where it is attached at the aitchbone.

After splitting the breastbone, make an incision near the top of the abdominal wall to pull the skin outward with your fingers. Gravity will pull the intestines down toward the bottom of the chest cavity and leave room for you to insert the knife and your free hand. Grip the handle of the knife with the blade turned toward you. You will slice downward and cut with the heel of the blade and push the intestines away from the knife as you slice down the belly.

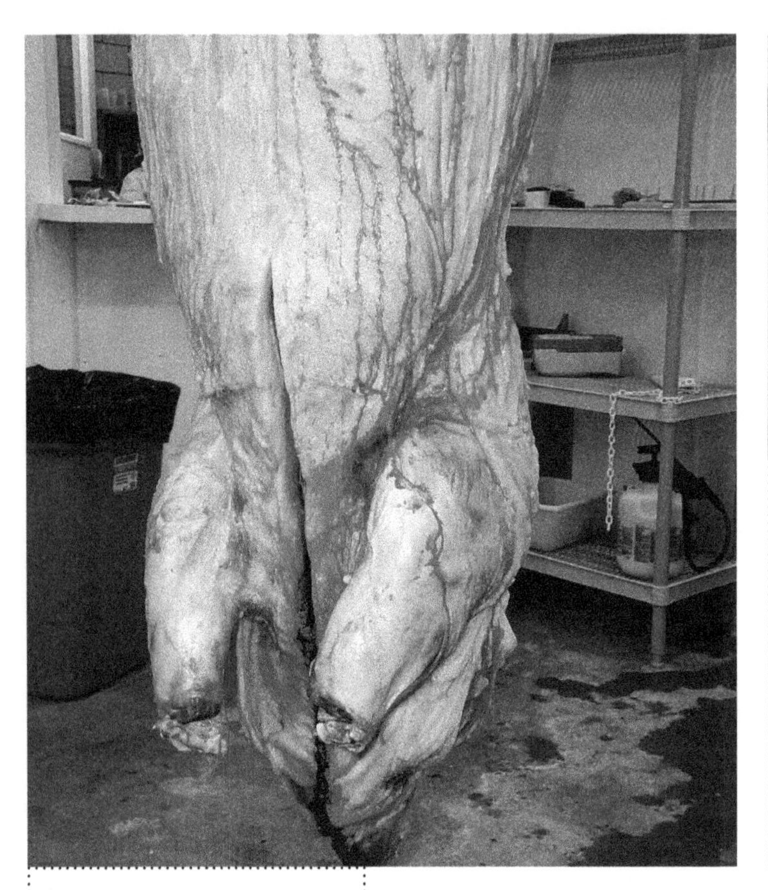

2. The sternum can be split while the carcass is on its back or after it has been lifted. This will allow access to the thoracic cavity once the abdominal wall is opened.

3. Open the abdomen by starting at a midline point at the pelvis. Once you carefully slice an opening large enough to insert your hand and knife, turn the blade outward while holding the heel with your hand on the inside. Use your hand to hold back the intestines as you slice down the midline in a smooth, continuous motion until reaching the opening of the thoracic cavity.

Keep the intestines away from the blade so that you don't cut them and spill their contents, which can contaminate the cavity. As you reach the parted breastbone, the intestines will fall forward and downward. They are still attached by muscle fiber and will not fall far. This is an easier method than drawing the knife upward to slice open the belly. Although it is a bit awkward, it minimizes the chance of puncturing the intestines.

The next step is to split the aitchbone. This will separate the hips and make the cut down the spine easier. A tub should be placed under the carcass to catch the viscera as you pull the kidneys, heart, liver, and stomach toward the opening. First, make a cut in the center between the two hams until you reach the aitchbone. This can be severed either with the heel of the blade or with a meat saw. At this point, the intestines are still suspended by the gut leading to the anus. Before you make a cut around this, tie the end of the anus

4. As you slice down, the viscera will fall down and out, but because they are held by connective tissue, they will not come out completely. Sever the connective tissues to allow the intestines and internal organs to fall free from the body cavity. Have a tub placed under the carcass before making this cut to catch the viscera.

securely with a cord to keep any fecal contents from falling out. Start in front and cut completely around the anus until it is free. This should allow the viscera to fall outward and downward. The diaphragm will now be exposed, and you will see the gullet that leads to the stomach. When you sever the gullet, the entire mass of viscera should come free and drop into the tub.

You can place the viscera on a table to cut off the liver and wash it in clean, cold water. Trim out the gallbladder and remove the spleen. The stomach should be tied off with a cord and cut free. The heart and lungs should still be inside the carcass cavity at this point, located in front of the diaphragm. Make an incision in the diaphragm where the red muscle joins the connective tissue. This will expose the heart and lungs, which should be pulled downward and cut free from the backbone. Trim any fat off the heart and lungs and wash them with cold water.

If you plan to use the intestines for sausage casing, they will have to be cleaned and rinsed with a salt solution several times. The easiest way to do this is to turn them inside out after cutting them into several lengths and scrape off the mucous coating. Generally, the small intestines are used for sausages, so tie off the large intestine and sever it from the small intestines and discard it. The heart and liver can be saved and ground in with meat for sausage.

## SPLITTING THE BACKBONE

The hanging carcass should be split apart while it is still warm. First, wash the inside of the carcass cavity and, using either a hand or electric meat saw, slice down the center of the backbone. Be sure to make a straight cut, or you may damage some of the loin areas. You can leave about 12 inches (30 cm) of skin uncut at the shoulders to keep the carcass from separating if you are concerned about it slipping off the gambrel. If you are not concerned about it slipping, continue to separate the back.

If you choose, the hams can be partially filleted while the carcass is suspended. Start your cut at the flank and continue to follow the curvature of the ham until you reach the pelvis; do the same on the other side.

## CHILLING

Your carcass is now ready to be chilled. A cold carcass is easier to trim and cut up than a warm one. Cooling it quickly also minimizes bacterial growth and souring of the meat. It is easier to cool the carcass when it has been split apart, as the air circulates around more of the body.

To properly chill your carcass, have a separate tub or vat large enough to completely submerse both halves in ice water maintained at a temperature between 34°F to 38°F (1°C to 3°C) for a minimum period of 24 hours. If you are using a refrigerator, it is possible to maintain a temperature of 38°F (3°C) at the bone within 12 to 24 hours. Chilled carcasses should not be worked with until all the tissue heat is gone. When it is thoroughly chilled, you are ready to cut up the carcass.

## CUTTING THE CARCASS

Before you begin to cut up the carcass, make sure you have sharp knives and several tubs available with cold water mixed with salt to start the curing process. One cup (288 g) of salt for 2 gallons (7.5 L) of water is a good mixture. There are many ways of creating salt brines for curing meat. Too little salt may result in spoilage, while too much salt creates hard, dry, over-salty meat.

If you use a meat saw to make the cuts, you should scrape the bone dust from the cuts after sawing. This mixture of small particles of meat and bone results from sawing. Cleaning off the saw blades or the cuts makes them less "crunchy" and reduces the chances of creating a bacterial haven.

Place one side of the carcass on a clean table, and start by removing the front and rear feet. Using the meat saw, cut off the legs at the hocks and the knees. The hind feet are generally not used because they contain a very high proportion of bone to edible meat. However, the front feet have a larger percentage of muscle to bone and can be used as pickled pigs' feet or trimmed and used for sausage.

The next step is to remove the ham. Begin with a cut at a point about $2^{1}/_{2}$ inches (6.5 cm) in front of the tip of the aitchbone and then cut through the fifth and sixth lumbar vertebrae. After the bone has been severed with a saw, use a knife to complete the cut through the rest of the tissue. Trim off most of the fat, but leave about $^{1}/_{4}$ inch (6 mm) on the whole ham. The pelvic bone will still be part of the ham and may be a problem in packaging or cooking because of its large size. You can trim out the bone and cut the ham into smaller pieces for easier cooking and packaging. You can also make bone-in roasts by cutting across the face of the ham to create ham steaks.

The next step is to saw off the shoulder at the third rib, counting from the neck. The shoulder has two major primal cuts—the Boston butt and the picnic shoulder—and three minor cuts—the neck bones, the jowl, and the clear plate. The shoulder can be kept whole, cured, smoked, or it can be divided into Boston butt and pork shoulder picnic.

The first step in cutting the shoulder is to remove the neck bones. There will be seven neck vertebrae, regardless of the length of neck. Trim these out as completely as possible. The neck bones can be used for soup stock or sauces, or may be barbequed.

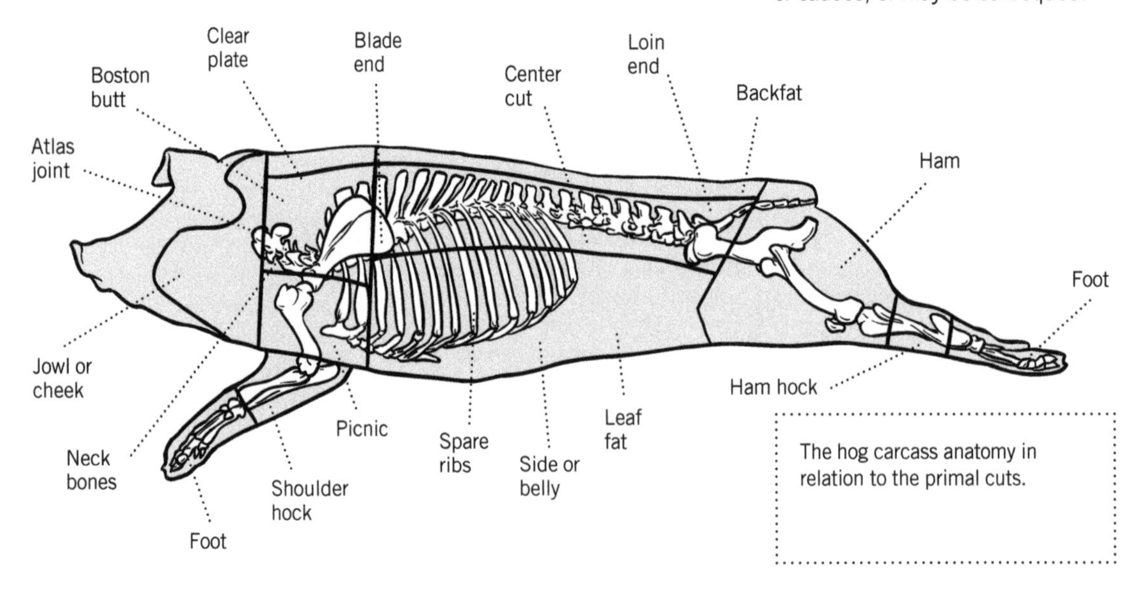


The hog carcass anatomy in relation to the primal cuts.

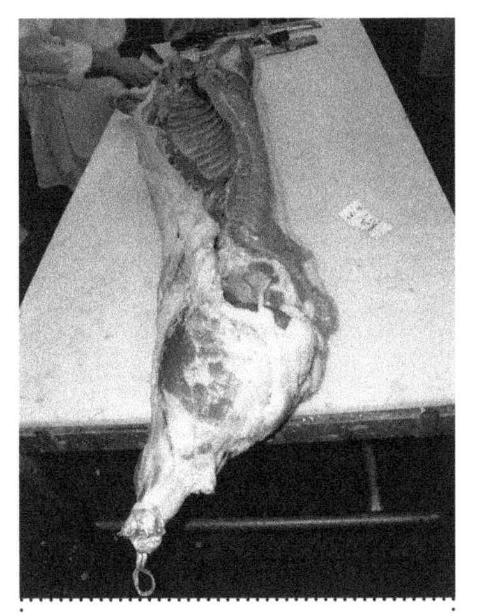

1. To begin fabrication, place one carcass side on a clean table. There are four major cuts to be made to separate the ham, shoulder, loin, and belly. To remove the ham, make a cut perpendicular to the leg bone from 1/2 to 2 1/2 inches (1.5 to 6.5 cm) anterior to the aitchbone.

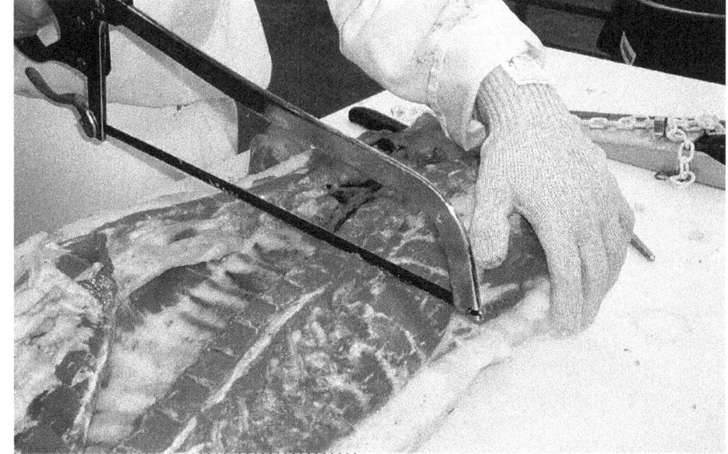

2. Separate the shoulder from the loin by sawing between the second and third rib, perpendicular to the back. This will separate the Boston butt and picnic shoulder from the belly and loin.

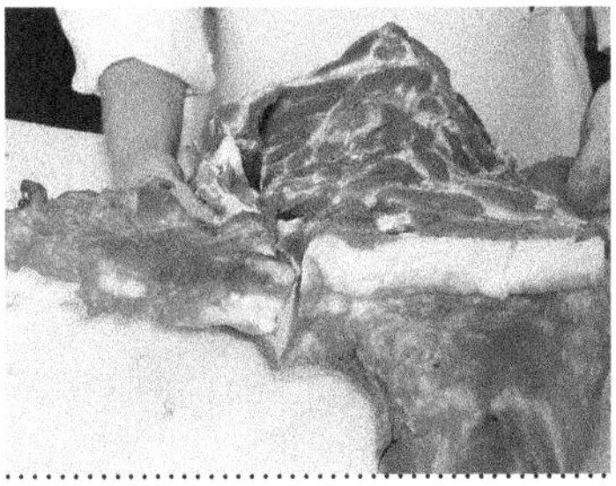

3. Removing the jowl begins at the fat collar immediately above the foreshank and continues straight across the top part of the shoulder. This should be trimmed of muscle and the fat set aside for sausage making.

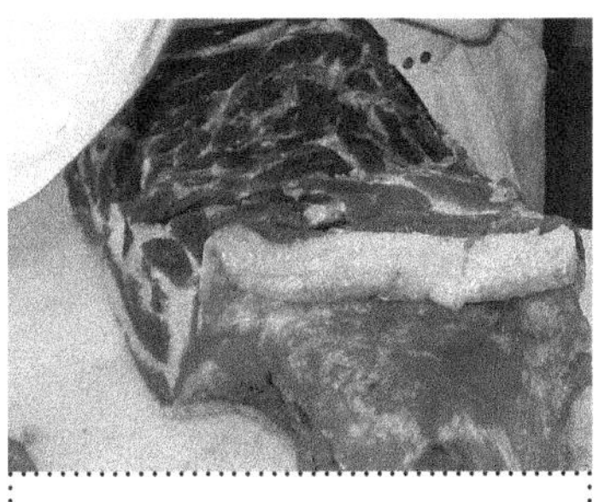

4. After removing the jowl, remove the clear plate, a fat cut much like the backfat, from the shoulder. The shoulder is composed of two wholesale cuts, the Boston butt and the picnic shoulder.

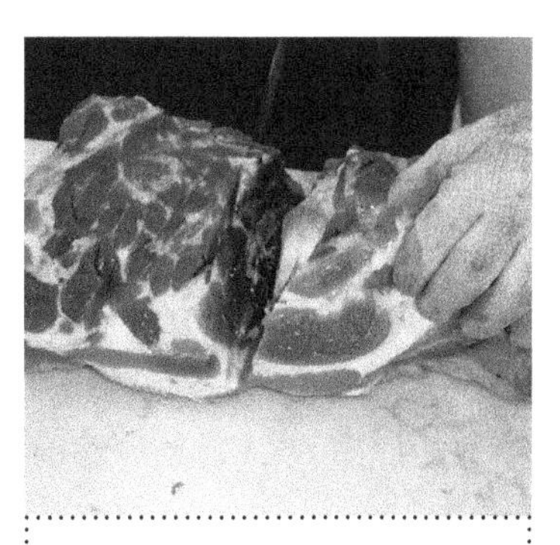

5. Separate the Boston butt and picnic shoulder by cutting 1 inch (2.5 cm) below the shoulder blade toward the leg and parallel with the sternum. Make the first cut with a knife, and then with a saw to sever the blade bone. Continue to trim excess fat down to 1/4 inch (6 mm) or less when making your cuts.

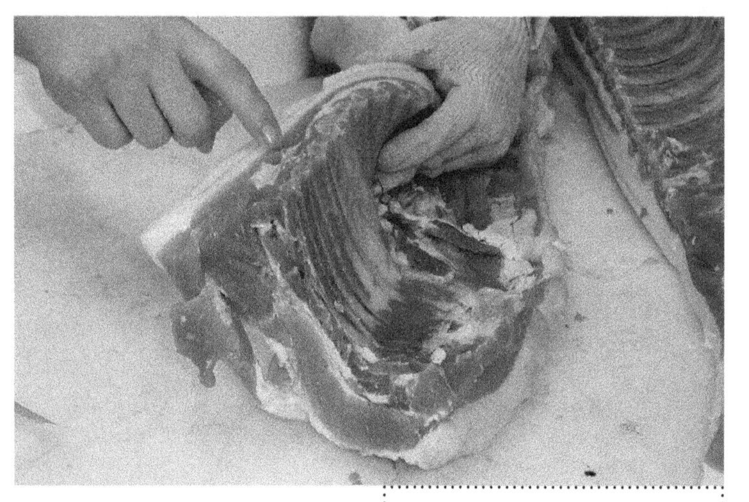

6. Separate the loin from the side (belly) by making a long, straight cut from the first rib (anterior) close to the backbone to the ham end, where the cut will be next to and closely follow the tenderloin, without cutting into or scoring the tenderloin.

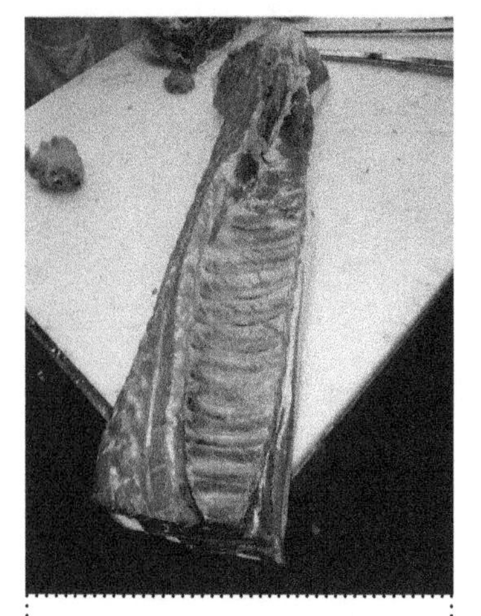

7. The whole pork loin is comprised of a blade section, a center section, and a sirloin section. The whole loin can be cut into bone-in pork chops or roasts. Or if preferred, the bones can be removed to make boneless pork chops or loin roasts.

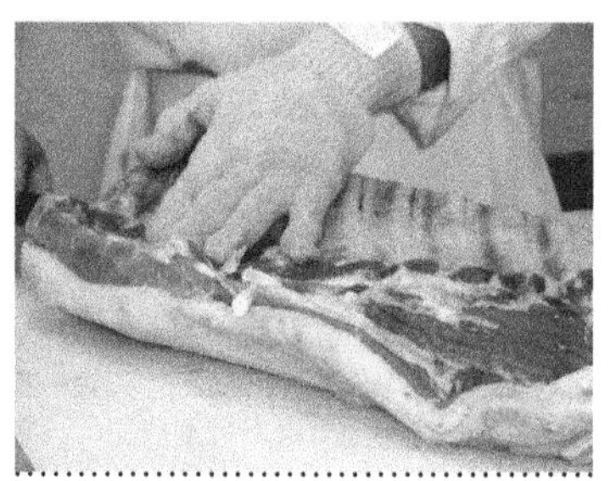

8. Bacon is made from the belly after the spare ribs have been removed. Square up the belly, trim the fat, and remove any rudimentary mammary glands and teat lines. It is then ready for smoking and curing.

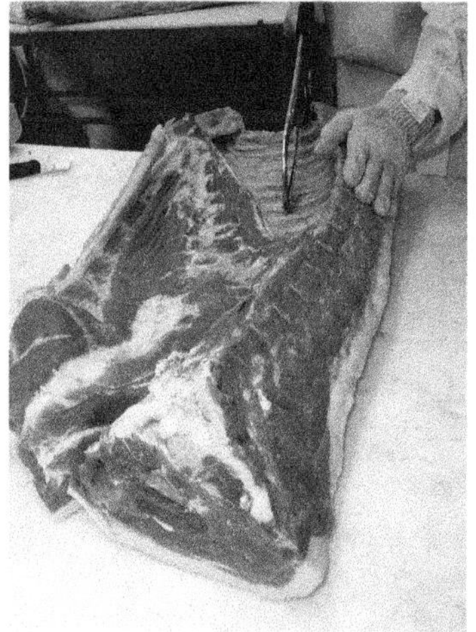

9. The spareribs are removed from the belly after the belly is separated from the loin. Be sure to trim out all bone and cartilage from the belly, as this is not desirable in bacon.

To remove the jowl, cut at the fat immediately above the foreshank and continue across the top portion of the shoulder. Trim out as much of the muscle as possible. This piece can be smoked or used in sausage.

The clear plate is a fat cut, much like that of backfat. It is removed from the top part of the shoulder by trimming close to the Boston butt. This large fat piece can be trimmed of any lean and the rest discarded or rendered, if you choose.

Divide the shoulder into picnic and Boston butt by cutting about 1 inch (2.5 cm) below the shoulder blade and parallel with the breast. The most popular cut of the Boston butt is the pork shoulder blade steak, which contains only one blade bone. Square the picnic by sawing off the foreleg. Most bones in the shoulder are located in the picnic, including the foreshank bone and the arm bone. First remove the foreshank, which is high in connective tissue, before removing the arm bone.

The loin is usually the most valuable cut in the pork carcass and may be about 16 percent of the carcass weight. The pork loin is a longer area of the carcass than that of the beef or lamb loin. The loin is separated from the shoulder by sawing across the third rib.

To separate the loin from the ribs, make a straight cut from a point close to the lower edge of the backbone at the shoulder to a point just below the tenderloin muscle from which the ham was cut. The spareribs and belly are now separated from the loin. When trimming the loin, leave about 1/4 inch (6 mm) of fat. The trimmed pork loin has a center that is higher valued than the two ends, which include muscles from the leg or ham and shoulder.

Begin your cuts with the end of the loin that was nearest the shoulder. Make cuts between each rib bone and the attached cartilage to create blade chops until you reach the fifth rib. The loin from the fifth to the tenth ribs will yield the center-cut loin chops, which are very desirable because they contain the tenderloin. The cuts from the rear end of the loin are referred to as sirloin chops, which contain portions of the eye and tenderloin, the top sirloin muscle, and hipbone.

The portion left is the belly or side. This will contain parts of the ribs that must be trimmed. You can use a straight knife and cut between the backfat and the belly to remove the ribs. The length of the ribs and the width of the belly will be determined by the location of the cut you made to separate the loin. You should remove all remaining bones and cartilage with the spareribs, because these will make very unpalatable bacon.

Once the spareribs have been removed, you can slice the belly or side into strips for bacon. You should square up the belly by trimming the outside parts evenly. This will remove any rudimentary mammary glands and teat lines that remain. The most common use of the belly is to cure and smoke it for bacon. Some people enjoy fresh pork belly, also called side pork, which may also be sliced. Most bellies are skinned before being cured and smoked.

The spareribs may be cured and smoked but can also be used fresh and barbequed. Generally, spareribs are cut into portions containing between two to six ribs.

CHAPTER 6

# POULTRY AND OTHER FOWL

**BUTCHERING POULTRY AND OTHER FOWL HAS BEEN A HUMAN EXPERIENCE FOR THOUSANDS OF YEARS. THROUGHOUT THE WORLD, CHICKENS ARE COMMON LIVESTOCK IN AGRARIAN CULTURES. THE EGGS AND MEAT THEY PROVIDE ARE AN IMPORTANT SOURCE OF PROTEIN.**

Start with healthy chickens for home butchering, either by raising them yourself or purchasing them from other growers.

Withhold feed for eight to twelve hours before butchering, but allow full access to water. A well-grown, 7-pound (3 kg) bird can have a dressing yield of 70 percent, giving you a 5-pound (2 kg) carcass.

As recently as 70 years ago, it was still very common on U.S. farms for chickens to be butchered at home. Providing meals for large gatherings, such as wedding dinners or for crews harvesting crops, meant chickens were butchered in the morning and served later that day. Today, refrigeration allows at-home butchering to be done anytime and the poultry or fowl are frozen for use throughout the year.

In the United States overall consumption of chicken has greatly increased in the last 10 years. Several reasons account for this increased consumption including lower fat content of the meat, lower store price, and the versatility of use in meat dishes.

Many cities and villages have adopted ordinances to allow raising backyard poultry. This is most often for egg production; however, once the hens reach the end of their productive egg-laying life, they are generally culled and can be butchered. By understanding safe handling procedures for chicken and fowl, you can harvest them any time.

Chickens may be kept in their coop until needed, or you can move them into a more confined area where they may be easier to catch. Depending on the number of people helping, you may want to divide the work into equal sessions. For example, start with four birds and complete all the steps from head removal to ice chest before beginning work on a second group.

## RAISING POULTRY

Raising poultry and other domesticated fowl is relatively easy. You will need to provide the basics of food, shelter, and water, but, by and large, poultry are very self-sufficient. Many books and guides are available on how to raise poultry and other fowl, including Quarry Books' *The Chicken Whisperer's Guide to Keeping Chickens*, and you should consult them.

## PURCHASING BIRDS

If you don't wish to raise them yourself, you may be able to purchase birds for butchering from local farmers. This will reduce your feeding and housing needs, but not necessarily be less costly than raising them yourself. If its flavor and texture you're looking for, then the cost of raising them will seem minimal. If purchasing birds to butcher, make sure they are healthy and free of any physical defects, such as damaged limbs, wings, or skin tears. Breaks or bruises will have some impact on the quality of muscle you will harvest as meat. Healthy birds will give you healthy meat; unhealthy birds will not.

You may be able to buy whole dressed birds and finish cutting up the birds yourself. If purchasing this way, keep the carcasses cold during transport and until you are ready to cut them up. Time and temperature will be your two biggest allies in processing your chickens, but they can also be your two biggest concerns in regard to food safety.

Assemble your equipment and lay it out before you begin. Use a sturdy table for cutting up the carcass. Supplies should include soap; pans for icing internal body parts, such as the heart, liver, and gizzard; clean water; towels; and any other item you may want handy.

# BUTCHERING BASICS

There is more than one way to butcher chickens, and you can make the process as elaborate as you wish, but a simple procedure is explained here. Processing poultry requires four basic steps, which should be done in separate areas to prevent contamination:

1. Slaughtering
2. Scalding and plucking feathers
3. Eviscerating (cutting open and removing the internal organs) and washing
4. Chilling and packaging

Arrange your work area prior to starting to help move the process along swiftly and safely. If properly done, your processing can be a pleasant experience.

# EQUIPMENT NEEDED

The tools, knives, and equipment needed for processing a few birds is often less than if you are handling many birds, although the processing principles are the same.

In the most basic operation, you will need knives for eviscerating and cutting, an axe and chopping block for removing the heads, several five-gallon (19 L) pails, a scalding tub, heating coil, a propane tank, a canvas or tarpaulin, and a sturdy table. Your chopping block will work best if you pound two large nails into it at distance of about 1 or 2 inches (2.5 to 5 cm) apart, depending on the size of the birds. For ducks and geese, you may want to increase that width an inch (2.5 cm).

The area you use for processing should be clean, have plenty of water available, and be as free from flies and insects as possible. Working early in the morning is often a good idea if you expect flies and insects to be a problem later in the day. Scrub tables with soap, water, and a diluted chlorine solution prior to use. If this is not possible, use a disposable plastic cover.

Sharpen and sanitize all knives before starting. Keep in a clean and accessible area.

You can use galvanized or plastic garbage cans or pails to hold the cooling water. Be sure these containers have been thoroughly washed, sanitized, and rinsed with clean water before they are refilled with carcasses to be chilled. Set up similar cans, pails, or plastic-lined boxes to use for feathers and unwanted body parts. While butchering, keep a water thermometer handy for checking the scalding water, which should be kept between 120°F to 160°F (49°C to 71°C).

Use sharp knives for working with the carcasses. These can include a straight boning knife shown on the left, a meat shears for cutting cartilage and bone, a butchering knife, a steel for sharpening knives during use, and a small, folding knife. Other knives can be used as needed.

Other equipment or supplies may include 5-gallon (19 L) pails for bleeding the birds. Using a tarpaulin on which you can pluck feathers will make for easier cleaning later. The feathers can be composted or placed in a plastic-lined bin for disposal.

Keep your packing materials close by. Have plastic-lined boxes or portable coolers filled with ice where you can cool the eviscerated carcasses quickly.

Make a list of all things needing to be done. This will allow you to envision the process from beginning to end. It is better to make adjustments at this stage than when the processing begins. Identify areas where the potential for contamination may exist, and keep these in mind as you are working so you can avoid them. Once all your equipment is clean and set out, you are ready to begin.

# HANDLING AND SLAUGHTERING

The chickens or other fowl you choose for slaughter should be taken off feed between 8 to 12 hours beforehand. Always provide them access to water. The removal of feed allows time for the crop and digestive tract to empty, helping prevent contamination during butchering. Birds can be kept in the same area used for housing them and then moved into a smaller, more accessible enclosure when ready to begin.

The method you use to slaughter the bird may involve an axe, a killing cone, or suspending them by their feet with a shackle or cord. Using an axe will require some coordination with a heavy blade, wood block, and an agitated bird that may not wish to cooperate. Safety with a sharp axe blade is very important, and it may help to restrain or tie the feet and legs of the bird together. If you feel you can hold the feet and legs steady while chopping off the head, you may not need to use any other restraints. However, several other methods can be used, such as wrapping the body in a linen sack or cloth. This will hold the wings close to the body and immobilize them. Once the head is chopped off, you will need to hold the bird by the legs for a few moments until its reflexes stop.

A cone is a safe and effective aid in restraining the bird for butchering. It is a large funnel in which the bird is placed upside down with its head falling through the small opening at the bottom. This method accomplishes several objectives by restraining the bird safely, preventing damage to its body, and using the downward pressure of its weight to force the blood toward its head to aid in bleeding.

1. The butchering process proceeds quickly once the head is removed. Before you begin, make sure the water temperature in the scalding tank is between 120°F to 160°F (49°C to 71°C). A simple method is to set up an open flame heater fueled by bottled propane that can be regulated. Plucking feathers will be easier if the water is maintained at a constant temperature during the entire butchering session.

2. You can use a wood block with two nails set 1 to 2 inches (2.5 to 5 cm) apart to chop off the heads of your chickens. A sharp hatchet will make a swift, efficient cut to remove the head.

3. Begin by grasping the chicken's feet and hold it upside down to immobilize it until you place its head on the block. Stretch the neck by pulling on the legs as the head is held between the nails. One swift chop should be sufficient to sever the head from the body. Be sure to keep your hand and fingers clear of the hatchet.

Shackles or cords can be used to hold the feet steady and keep the legs apart as the bird is suspended upside down. A clothesline or other sturdy design can be used to hold a bird off the ground and will allow you unencumbered movements around it. Suspending the bird at your eye level will help with the process. If using this method, allow time for the bird to settle down before beginning.

Whether using an axe, cone, or shackles, you want to have a clean and humane kill. With an axe, you can chop against a wood block with one quick motion. Hold the bird tightly by the feet and legs with your free hand. Place its head on the wood block between the nails and stretch out the neck by pulling back on the legs. One quick chop with the axe should be sufficient.

If using either a cone or shackles, you have ready access to the neck of the bird. There are several options for killing the animal. The simplest is to sever the head completely. Another option is to sever the jugular veins. This is done by making a cut just behind the jaw. This cut should sever the veins without cutting the esophagus or windpipe. Cutting only the neck vein, reduces the chance of carcass contamination by blood being drawn into the air sacs. This is a humane method because the bird is unconscious due to the loss of blood from the brain.

To do this, hold the beak with one hand and pull down slightly to steady the bird. There are two veins in the neck leading to the head, and both pass near an ear lobe. Be sure to hold the front part of the head firmly to avoid cutting your hand. Press the point of the knife into the flesh, lift the handle upward, and cut downward with the blade severing the veins. This should result in a good bleed. If not, try again until there is free bleeding.

Once the head is removed, hold the bird upside down in a 5-gallon (19 L) pail or similar container for 15 seconds to begin the bleed. The bird's reflexes will continue to flap the wings, but the confinement will eliminate or greatly reduce any damage to the bird. They can remain in the pail or container until you are ready to scald the carcass.

## BLEEDING THE BIRD

It will not take long to bleed the bird. However, it is still important that your bird is killed in a manner that allows as much of the blood to drain from the body as possible while preventing damage to the carcass. Only about 50 percent of the blood is actually removed from a bird. What remains does no harm if the carcass is to be cooked immediately. Since blood spoils more quickly than other parts, it is beneficial to remove as much blood as possible to lengthen the shelf life for either fresh or frozen poultry or fowl.

Be aware that any method that involves beheading or breaking the neck will accomplish the killing but will not produce the same type of bleeding as severing the jugular veins because the heart stops when the spinal cord is severed.

If you are removing the heads, you can bleed the birds by placing them upside down in a 5-gallon (19 L) pail once the head is off. Hold them by their legs until their reflexes stop to prevent any damage to their body and leave them in the pail until you are ready to scald them.

If using a cone or shackles, you can let the blood drop into a container below for easier cleanup. When the bleeding is finished, you are ready to remove the feathers.

# SCALDING CARCASSES

Birds must be properly bled and all body reflex movements stopped before any scalding should be done. Hot scalding, with water temperatures above 155°F (68°C), is an easy, quick method to remove feathers. Start by holding the bird tightly by its legs and immerse it neck first into the scalding water. It is important to get enough water into the feathers. Move the bird up and down and from side to side to get an even and thorough scalding, which will make the feathers easier to remove. Repeated dips may be needed, but don't overdo it to prevent burning.

One simple rule to follow when scalding is that the higher the temperature of the water, the less time of immersion needed (although another method that involves immersing the carcass for longer periods at lower temperatures can be successfully used too). You can avoid over-scalding by following the temperature and time recommendation for the birds you are using. Over-scalding causes the skin to tear and discolor and gives the bird a cooked appearance; the carcass will lack bloom and turn brown rapidly, or bright red when frozen.

Water that is hot will cause the outer cuticle layer of the skin to slough off as the feathers are plucked from the carcass. This cuticle layer is the yellow pigment area commonly seen on dressed chickens. The use of high temperature for a shorter period of time, while it increases the ease of plucking, risks the loss of this yellow cuticle layer of the skin, which may result in the skin tearing more easily. If you choose not to keep the skin, this may not be a concern.

For young birds with tender skins, the scalding temperature should be between 125°F to 130°F (52°C to 54°C) for 30 to 75 seconds. It may reach between 155°F to 160°F (68°C to 71°C) for older birds for the same length of time. At this temperature, the cuticle covering the skin typically will be removed.

Feathers from waterfowl, ducks, and geese are more difficult to remove. Scald these birds at higher temperatures of between 160°F to 170°F (71°C to 77°C) for 1 to 2 minutes. Waterfowl have natural water repellant oils in their feathers. You can add detergent to the scalding water used on waterfowl to help the water penetrate through the feathers.

To scald the carcass, hold the feet and gently dip it into the hot water. Hold for five seconds before pulling out. Then, dip again while slowly moving it from side to side, completely submerging the feathers. Do not overscald the bird.

Fat birds will hold their color longer because the melted fat forms a film over the skin, reducing the effects of the air. You can increase the yellow coloring on fat birds of the yellow-skinned variety by dipping them into boiling water and then immediately plunging them into cold water. The hot water melts the fat and draws it and the yellow pigment to the surface of the skin. The cold water causes the fat to harden and the color to set in the fat.

1. When scalding is complete, begin to remove the feathers by pulling on them. You can place the feathers on the tarpaulin for later disposal. The large feathers can be removed quickly. If needed, you can quickly dip the carcass again to loosen the rest.

2. Some pin feathers may remain after plucking. There are several ways to remove them if you choose to keep the skin on, including singeing, removal with a small knife, or continued plucking by hand. Pin feathers can remain on the carcass through cooking with no effect upon the meat. Whether you keep the pin feathers completely intact is a matter of personal choice.

## PLUCKING FEATHERS

Picking off the feathers, or plucking feathers, is the next step. Birds should be plucked immediately after scalding. You can lay them on a canvas or tarpaulin, or suspend them by their legs to do this. There is no one correct way to remove the feathers, and all feathers need to be removed.

You can remove the tail and wing feathers first and then the rest of the body feathers. If properly scalded, the tail and wing feathers can be quickly and easily removed before you move on to the main body. Chickens and other domestic and wild fowl have pinfeathers, which are tiny, immature feathers lying below the surface of the larger feathers. They are more difficult to remove because of their size. Remove them by using a pinning knife or dull knife to gently scrape or pluck them off.

Plucking feathers is not hard but takes time and patience. Work quickly and dip the carcass again if needed to avoid it drying out. You can use a rolling or rocking motion to remove feathers or pull in the direction they grow to minimize skin tears. If you intend to show your carcass at an exhibition, tears in the skin need to be avoided because the skin needs to be completely intact.

Inspect the carcass to ensure all feathers have been removed. If some very fine "hairs" remain, you can remove them by a process called singeing. This is where a gas bottle torch or an open flame on a gas range can be used to burn them off. The potential for injury to you or your bird by using this method may not outweigh the advantages of removing every last one, particularly if you don't intend to eat the skin. Singeing is usually not necessary on young birds, but more mature chickens and turkeys may have hairs remaining after the feathers are removed.

# DRY-PLUCKING FEATHERS

Dry-plucking (also called dry-picking) is a process used to remove feathers from birds such as waterfowl. This involves removing the feathers without first immersing the carcass in water. However, this method requires a prior process called debraining, which relaxes the feather muscles, aiding in their removal. This process also requires the bird be suspended so that you can work more easily.

To begin, locate the slit in the roof of the mouth and insert a small-blade knife, blade edge up, into the cleft in the roof of the mouth at a slight angle. Force it toward the back of the brain with the handle about parallel to the upper beak. If properly debrained, the bird may give out a peculiar squawk. In contrast, a turkey will relax its wings and spread out its main tail feathers in the shape of a fan.

The puncturing of the brain relaxes the feather muscles, causing the feathers to become loose and more easily plucked. However, this condition generally only lasts about three minutes before the muscles begin to tighten up. You will need to quickly pluck the feathers with this method.

In dry-picking, it is easiest to remove the feathers in the order in which the parts of the bird bled out, beginning with the tail because it was bled upside down. Twist out the tail and the main wing feathers first, and then pluck the breast, neck, back, thighs, and legs. After the large feathers have been removed, you can begin removing the pinfeathers. Again, handle and pluck the bird so that the outside layer of skin is free from tears, bruise spots, or abrasions and cuts.

After you have completed plucking the feathers, rinse the carcass with clean water to remove any loose feathers, dirt, blood, or other foreign matter that may still adhere to it. You are now ready to dress the carcass.

## WAX PICKING

Wax picking works well to remove small feathers and down from ducks and geese after most feathers have been removed and the carcass has dried for a short time. Paraffin wax can be heated in a tub separate from the scalding water to about 135°F to 160°F (57°C to 71°C) to create a liquid bath. Dip the bird into the wax bath for 30 to 60 seconds, and then dip it into cold water to set the wax. The wax will adhere to the dry feathers, down, and stubs, which are very short broken feathers. You may need to dip the bird a second time if enough wax does not cover the bird.

While the wax is still flexible, you can begin to peel it off. This will remove any feathers, pinfeathers, hair, and down that has adhered to the wax. Finally, rinse or wash the carcass to remove any remaining particles and to moisten the carcass again.

## CHILLING THE CARCASS

Unless you immediately proceed to cut up the carcass, removing the body heat is important at this stage. Put the carcasses in a cold water bath with temperatures between 32°F to 36°F (0°C to 2°C). Birds should never be frozen before being chilled down because the meat will be less tender later as the muscle fibers slide and lock together. Placing the birds in chopped ice will rapidly cool them.

## EVISCERATION

To make your work easier, there is a proper order for evisceration, which is cutting open and removing the internal organs from the body cavity, plus the removal of the head and feet.

First, remove the head and neck. The head will have been removed if you used an axe and chopping block earlier. If the bird was suspended by its feet and you simply cut its throat, you will now need to remove the head and neck.

To remove the head, cut between it and the first neck vertebra, giving it a little twist as you cut. Avoid cutting through the spine. As you cut through the back of the skin on the neck, peel down the skin and sever the skin close to the shoulders where it enters the body cavity. This will expose the crop, trachea (windpipe), and gullet (esophagus). These can be removed by hooking the short gullet with your index finger and peeling the crop loose from the skin by pulling it out without using a knife. Next, cut off the neck by cutting into the neck muscle at the shoulder and then twisting it off. You can wash it and set it in a chilling pan for later.

Next, remove the shanks. To remove the feet, place the bird breast up on a table or a stable cutting surface while holding a shank (leg) in one hand. Cut through the hock joint by drawing the bottom part toward you as you cut away.

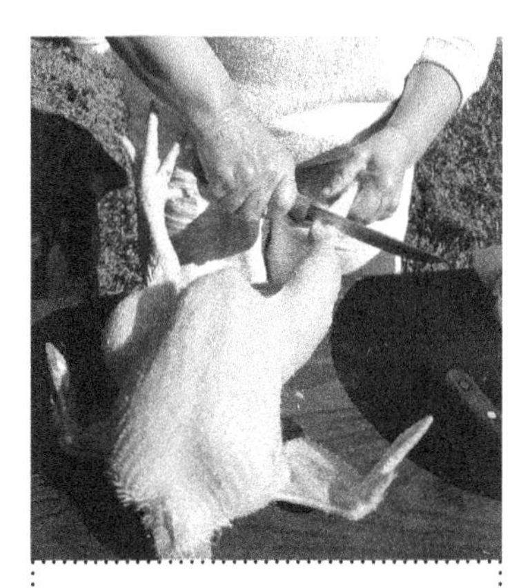

1. Begin cutting up the carcass by removing the feet. Make your cuts at the first joint. Feet can be discarded, or they can be used for soup stock.

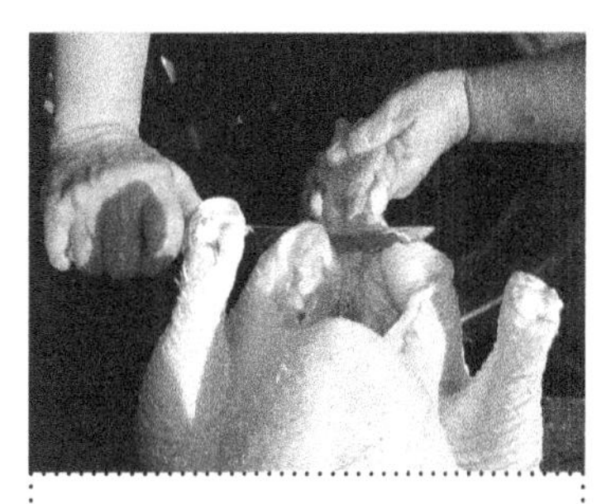

2. Next, place the carcass breast side up and remove the oil gland in the tail by making a first cut in the anterior portion of the rectum. Cut along the sides as you pull back on the tail. Sever the oil gland by making a final cut where the tail vertebra joins the backbone.

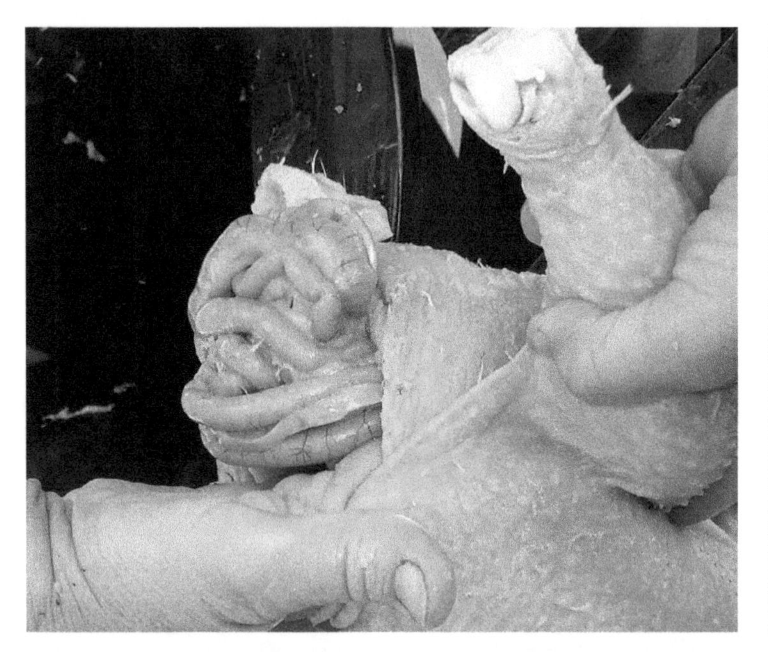

3. The opening created by removing the oil gland should be large enough for you to reach your hand into the body cavity pull out the internal organs and intestines. Make a short cut with your meat shears if the opening needs to be larger. Draw out the viscera and organs, and place them on a clean surface.

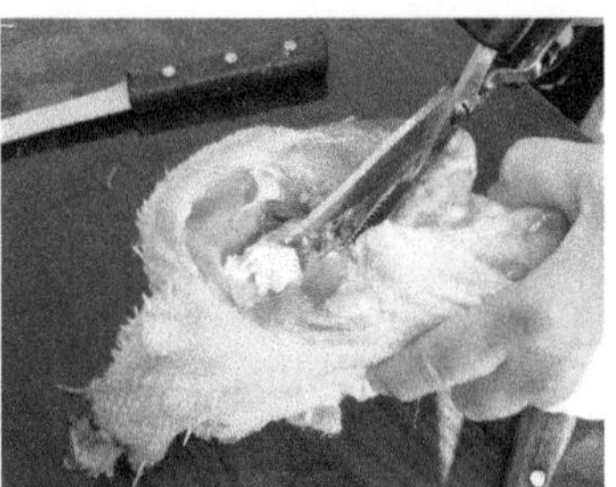

4. Open the body cavity using a meat shears, cutting the breastbone lengthwise to the neck if it is still intact. If not, the neck can be cut from the body now more easily than if done earlier. Then, use the shears to cut the bottom half of the body.

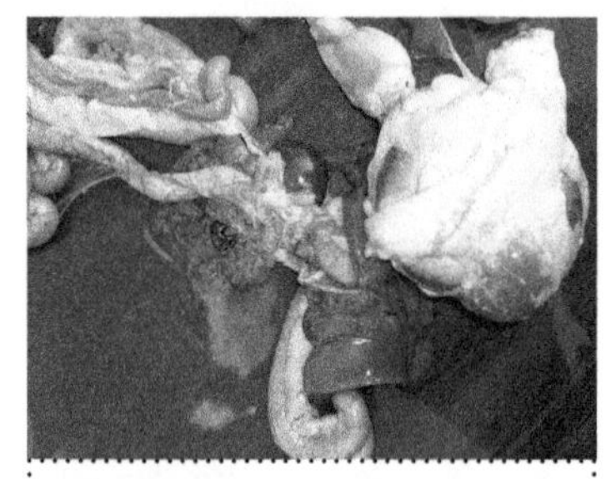

5. The heart, liver, lungs, gizzard, and other organs and viscera should have a healthy look. Examine them. If they appear off-color or any lesions are noticeable, it may indicate an unhealthy bird, and you may want to consider disposing of it. A healthy bird will have bright-colored, vibrant organs and viscera.

6. The gizzard is highly prized by some people. Slice it in half and remove any feed contents that may remain. It should be nearly empty if feed was withheld for 12 hours prior to butchering. The lining on the inside should be removed before cooking and is easily peeled off.

With the bird breast down on the table, you can remove the oil gland near the tail. The oil gland can be seen at the base of the tail. Make a forward cut 1 inch (2.5 cm) from it and cut deep into the tail vertebra, then follow it to the end of the tail in a scooping motion to remove the gland.

The body cavity can be opened by making a small cut near the rectum. Be careful not to cut the intestines or contaminate the carcass with fecal material. Two types of cuts can be used to make this opening, depending how you will use the bird. A midline, vertical, or "J" cut is often used for broilers and other small poultry not to be trussed when cooked. The traverse or bar cut can be used for turkeys, capons, or other large fowl.

To make a vertical or midline cut, pull the abdominal skin forward and up away from the tail. Start just to the right of the breastbone with your knife point and cut through the skin and body wall. Extend the cut to the tail alongside the vent (rectum). Go slow so that you do not cut the intestine. Cut entirely around the rectum as you slowly pull it and the end of the intestine out and away from the opening of the body cavity.

To make a bar cut, cut a half circle around the rectum next to the tail. Insert your index finger as a guide to make a complete circle to free the rectum taking care not to cut the intestine.

Next, draw the intestinal tract, the heart, lungs, and liver through this opening. You may insert your hand to assist in extracting these parts. You can loosen the lungs from the entrance of the shoulder with your hand if these organs do not easily pull out. When the lungs are loose, you can use a scooping motion with your fingers to bring out the rest of the viscera (internal organs).

After the viscera have been removed and placed on a clean table, you can remove the green gallbladder from the liver. It can either be cut or pinched off. The gizzard, liver, and heart should also be removed. Cut the gizzard from the intestines and stomach. It can be split lengthwise and the contents washed away. The lining inside the gizzard should be peeled away and can be easily removed by using your fingers.

Remove the heart and trim off the sac and heavy vessels around its top. Squeeze it to force out any remaining blood. Rinse the giblets (heart, gizzard) well and place all the parts in a pan of cool water.

Next, wash the inside of the carcass thoroughly with clean, cold water after you have finished removing the insides. The carcass is now ready for cooling.

## COOLING

After evisceration is complete, cool the carcass as soon as possible by placing it in a cold bath of clean water at a temperature of 35°F to 40°F (2°C to 4°C). Once cool, it is ready for cooking, freezing, or cutting it up.

If birds are to be frozen, the gizzard, heart, and liver can be wrapped in waxed paper and placed inside the body cavity. The birds can then be placed in a moistureproof and vaporproof bag and frozen. Birds can be shaped to give them a plump, attractive appearance. Birds for roasting should be trussed by using cord or wire that is drawn over the fore part of the breast, over the wings, and then crossed over the back. Then, bring the wire over the ends of the drumsticks and tie it tightly at the back of the rump.

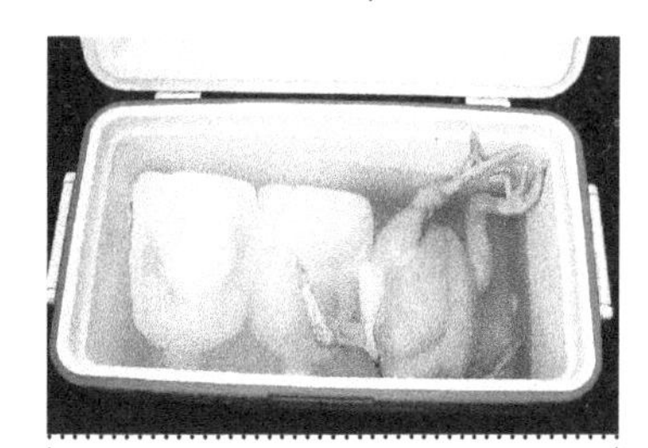

You may decide not to cut up the carcasses until all of the butchering, feather plucking, and viscera removal is completed for all the birds. If you wait to cut up the birds until later, you need to place the whole eviscerated carcasses in cold water to remove the body heat of each bird as quickly as possible.

# CHILLING AND PACKAGING

Before packaging a poultry carcass, it should be cooled to below 40°F (4°C) within two to six hours after cutting it up to maintain high quality meat and minimize bacterial growth. Small birds can be chilled in a couple of hours while turkeys, large capons, and roasting birds may require several hours to reach this temperature. If using a container with ice water, you may need to change the water several times. Always make sure that the water is clean and your container is large enough to submerge the entire carcass. For this you can use an ice water bath in an insulated chest, bucket, or other clean container.

# SKINNING

If you do not wish to use the skin later, you can remove it along with the feathers, saving you the scalding and feather plucking steps. The birds are killed and bled in the same way as those which have their feathers plucked.

To skin your bird, begin with a cut into the skin at the bottom of the breastbone with the carcass on its back and its head away from you. Lift the skin and cut it forward to the front of the neck. Peel the skin and feathers back with your hands and expose the breast muscles. Use your hands to work the skin loose from the thighs. Push back the skin to expose the hock joint and then cut through the joint. Remove the skin from this area on each foot. Next, loosen the skin to the joint between the first and second section of the wing. Then, remove the last two sections of the wing along with the skin.

Loosen the skin at the base of the neck and cut the meat around the base of the neck near the shoulders. Twist the neck off the carcass. The final cut is the removal of the tail and the attached skin with feathers. The carcass is now skinless, neckless, tailless, and only has the upper section of the wing left. Remove the skin with feathers and place the carcass in a pan of clean water while you clean and sanitize your table or cutting surface before eviscerating the bird.

# CUTTING UP BIRDS

Unless you decide to roast the entire bird, you will likely want to cut the carcass into various pieces. Most of these pieces are made by cuts at certain joints. Breaking down the bird into parts is a simple procedure using a sharp knife. Typically, the edible yield for fryers and broilers is about 65 percent with the rest lost as bones and viscera.

## LEGS, THIGHS, AND DRUMSTICKS

It is easier to cut up the carcass if you remove the wings and legs first. A leg includes the thigh and drumstick and is removed by making a first cut at the hip joint. The skin on the back or on the pelvic bone is not included with this cut.

Start by laying the carcass on its back. Cut the skin between the thighs and the body. Then, lift the carcass by holding a leg in each hand. Bend the legs back until the hip joints snap free. This will allow you to cut each leg off at the joint as close as possible to the backbone.

Next, cut through the knee joint to separate the thigh from the drumstick. If you are unsure of its location, you can find it by squeezing the thigh and drumstick together. The joint that moves is the one you are seeking.

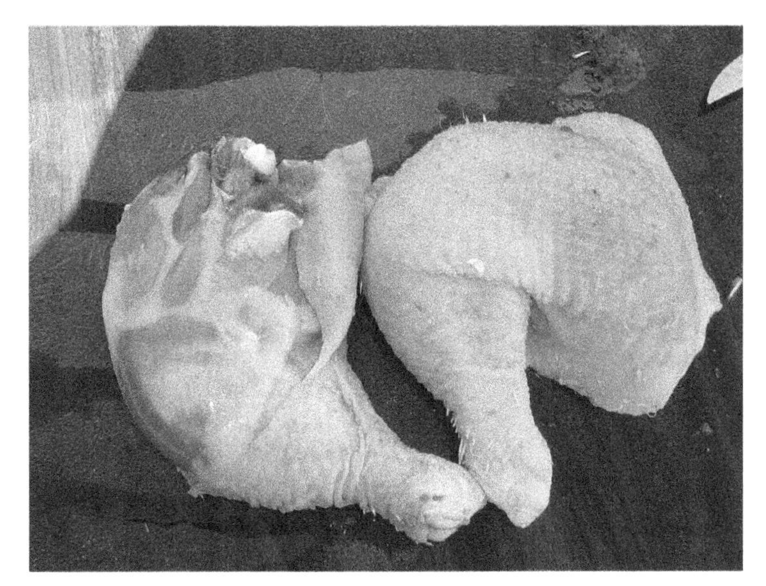

Your cuts should leave you with two legs and two breast pieces, as well as assorted internal parts. The heart and liver can be ground up to be used in sausage or dressing. After your cuts are finished, place those in cold water until you package them.

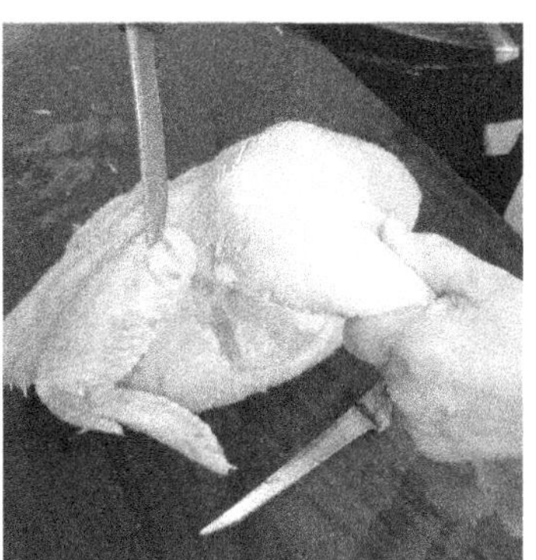

The wings should be removed by cutting as close to the shoulder as possible, severing them at the joints. Some prefer to remove the outermost tip of the wing while retaining the two inner shanks.

## WINGS

Wings include the entire wing with all muscle and skin intact. You can remove the wing tips, which can be used for soup stock. Cut through the joint closest to the body to remove them.

## TAIL

The tail can be removed by cutting along each side and through the joint at the end of the backbone.

## BREASTS

A breast is separated from the back of the bird at the shoulder. Start by placing the carcass on the neck end. Cut along the side of the backbone, starting from the tail and continuing through the rib joints to the neck. Then, bend the carcass back to find the joint before cutting through the meat and skin. The ribs may or may not be removed.

You can split the breast lengthwise by first placing it skin side down. Then, cut through the white cartilage at the V of the neck. You can bend each side back as you push up on the breast from the bottom to snap the breastbone free. The wishbone is the clavicle and can be removed by severing it from the breast. Make your cut halfway between the front of the backbone to a point where the clavicle joins the shoulder.

## TURKEYS

Domestic turkeys differ from wild turkeys, but they can be cut up in similar ways to chickens, only larger portions are involved. Butchering wild turkeys is discussed in a following chapter. A domestic turkey will yield a greater volume of meat than a chicken because it is a much larger and denser bird. A typical turkey will yield about 71 percent edible meat with about 29 percent of the carcass lost to bones and viscera.

Because of a turkey's large size, the containers used for scalding it need to be larger if you decide to keep the skin on. This also requires more water and a greater physical effort to handle a large, heavy bird.

## DUCKS AND GEESE

Waterfowl are difficult to scald because they are very tight-feathered. This allows them to stay dry and warm in their natural habitat by keeping the water and cold away from their bodies. But it presents challenges in butchering because you need to use a much higher water temperature or steam to help remove the feathers. To do this, wrap the bird in a burlap cloth and then immerse it in water that is near the boiling point. You can use a lower temperature to 160°F (71°C), but the scalding time will be much longer—up to two or three minutes.

You can dry-pick properly bled and debrained waterfowl, but you will need to remove the fine down feathers by scalding or waxing if you want to retain the skin.

Geese present a special challenge in removing all the pinfeathers. Typically, geese can be raised and ready for market within 10 weeks or less. To produce geese with minimal pinfeathers, they will require 20 weeks or more for them to mature.

## GUINEA FOWL AND PHEASANTS

You can dry-pick or semi-scald most guinea fowl as you do chickens, and they can be processed in a similar manner. Pheasants can be skinned or semi-scalded before you pick and eviscerate them.

## OSTRICHES AND EMUS

Ostriches and emus are flightless birds referred to as ratites and are distantly related. Both have the same genetic base as the turkey but reach a larger mature size with the ostrich and emu weighing up to 300 pounds (136 kg) and standing 6 feet (2 m) tall at the shoulder.

At one time ostrich and emu meat was a byproduct, as the birds were traditionally used for the production of feathers, high-quality leather, and fat. The feathers were used for making hats and long, fluffy scarves known as boas, while the hides were processed into expensive leather for shoes, boots, purses, and other accessories. Their 4-pound (2 kg) eggs could be sold to painters and collectors.

In recent years, their intense red-colored meat has appeared on gourmet menus in upscale restaurants and in-home processing. Their meat can be made into steaks, ground meat patties, sausages, and jerky. It is low in fat and cholesterol, and the best cuts come from the thigh and the larger muscles of the drum and lower leg.

If butchering one of these large birds, you should use good restraint methods for the feet and head, both of which can be used as weapons by a frightened bird.

Because of their large size, skinning the ostrich or emu would be the easiest method instead of trying to scald them. The feathers can be saved for other uses. Cutting up the ostrich or emu is roughly similar to cutting up a large turkey.

# GAME BIRDS

The major game birds taken by hunters include pheasants, quail, grouse, partridges, wild turkeys, and mourning doves. With the possible exception of the wild turkey, game birds are generally cleaned by removing the skin with the feathers intact because it is much easier and faster than plucking.

## PHEASANTS

Pheasants are popular game birds that are one of the meatiest for their size, commonly yielding a 1 to 2 pound (0.5 to 1 kg) carcass. The majority of their edible meat is located on the breast.

Pheasants grow in a number of states, and they are subject to a considerable variation of laws and regulations relating to the hunting season and the bag and the possession limits. Hen pheasants are off-limits for hunting in most states, and you should always check the regulations before hunting them. It is the hunter's obligation to know, understand, and abide by the state regulations. Because it is illegal to shoot a hen pheasant in most states, you are required to leave the head and feet on the bird while it is being transported.

You can clean pheasants using two different methods. The first is to skin the bird and process it in the same way as poultry. The second method is faster and neater and is preferred by hunters who do not like to handle the entrails. However, in states where the heads must be left on as they are transported across state lines, this procedure may not be legal.

In this second method, you skin the bird by making a small cut on the underside of the breast before pulling the skin and feathers off the carcass. Next, cut down both sides of the back, starting by the neck and cut through to the last rib. Pull the breast apart from the neck, back, and legs. This should leave the intestinal tract, heart, lungs, and liver attached to the back portion. Then, remove the feet and lower legs at the joints below the drumsticks. To remove each leg, cut through the joint attached to the back.

You will produce three pieces by dressing pheasants as two thighs and a drumstick and one breast. You can discard the back and neck because they contain very little meat, although they may be used for soup stock.

## QUAIL, GROUSE, PARTRIDGES, AND DOVES

Small game birds such as quail, grouse, partridges, and doves have the majority of their edible meat in the breast. This portion is the most used as the rest of the carcass has limited food value. Removing the feathers and skin together is a favored way of dressing them because it eliminates the tedious task of picking them off small bodies. It is generally easier to dress these birds by removing the entrails from the whole carcass.

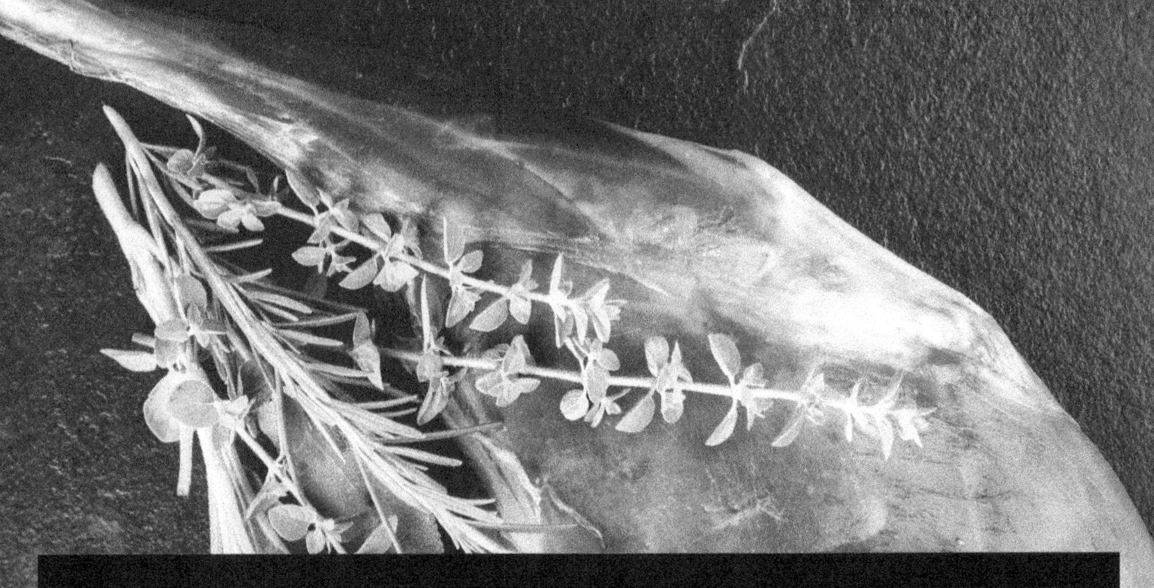

CHAPTER 7

# VENISON

**DEER ARE TYPICALLY REFERRED TO AS A LARGE GAME ANIMAL. YOUR ABILITY TO HUNT AND HARVEST THESE IS DEPENDENT ON THE STATE OR FEDERAL LEGISLATION THAT RESTRICTS THE PERIOD WHEN LARGE GAME ANIMALS MAY BE LEGALLY SHOT. CONSERVATION EFFORTS OF PAST DECADES HAVE PROVIDED FOR A MORE ABUNDANT POPULATION OF EACH.**

There are vast differences in size among big game animals, and these differences may influence your choice or your ability to handle a carcass in the field.

Large game animals, such as deer, may be hunted in the wild or sourced from commercial production facilities that specialize in raising big game. Regulations apply to both commercially produced and hunted game, though.

Large wild game animals have become an attractive addition to meal preparation for many families. Animals may be hunted in the wild or sourced from commercial production facilities that specialize in raising big game. Regulations apply to both commercially produced and hunted game, and if you choose to secure animals through either route, you are obligated to know and understand the rules and regulations governing them and to abide by them.

It is considered a privilege to hunt and receive licenses to secure game, so you are also obligated to minimize the suffering of the animal and to make a quick and efficient harvest to achieve optimal meat quality. You will also want to prevent or minimize waste of the carcass taken.

You should plan ahead before embarking on a big game harvest to ensure that you can safely and efficiently dress and transport the carcass, sometimes over large distances. Planning for as many variables as possible in the wild will reduce the risk to yourself and possible contamination of the carcass if you have a successful hunt. These may include transport, weather conditions, distance from shelter, and equipment.

## SAFETY

Without some precautions at the time of harvest, you may be at risk for injury or worse from the actions of a wild animal that suddenly comes back to life. You should approach all big game that has been shot with caution. Any wounded animal can turn aggressive, and those with antlers can be extremely dangerous for unsuspecting hunters.

## BLEEDING DEER

Having assured yourself that the animal is dead, you can consider bleeding it, although it is not necessary for deer unless it was shot in the head, and often not even then. Bleeding deer can be done by either cutting the jugular vein in the neck just behind the jaw or, in the case of a trophy-size buck, by cutting into the base of the neck several inches (18 cm) in front of the breast.

You can begin by inserting a hunting knife with a 5-inch (13 cm) blade into the breast with the point of the blade aimed at the tail. Insert the blade all the way and press it downward toward the backbone. With a slicing motion, withdraw the blade. Then, elevate the hind legs to allow gravity to drain the blood.

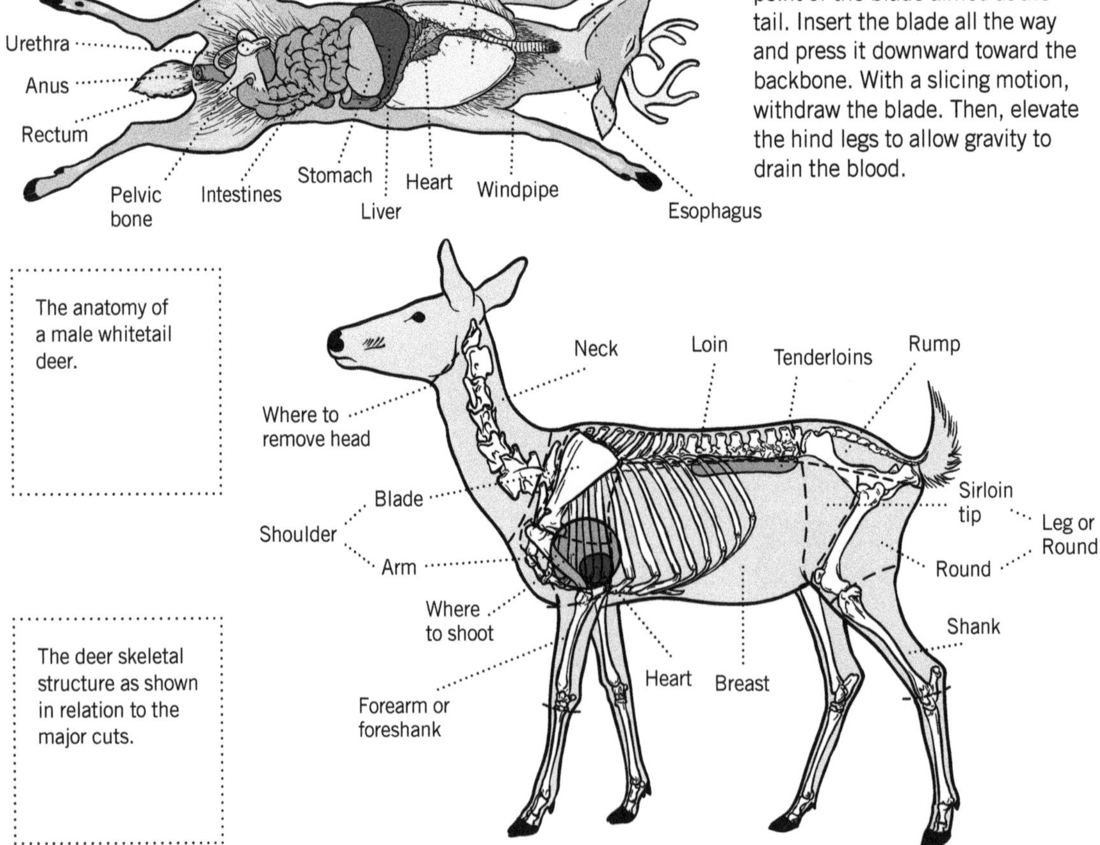


The anatomy of a male whitetail deer.

The deer skeletal structure as shown in relation to the major cuts.

# FIELD DRESSING

A clasp knife or a sheath knife work best for dressing deer in the field because they are easier to use than larger 10- and 12-inch (25.5 to 30 cm) knives. You can use small knives to make more precise and sensitive cuts if needed.

Temperature and insects are two variables that can affect the quality of any carcass you dress in the field. You can minimize the effects of flies by wrapping the dressed carcass in cheesecloth. This will prevent flies or insects from entering the body cavity and laying eggs. Warm weather can cause spoilage in a short time, and getting the carcass to a refrigerated cooler within a day of killing it will prevent or minimize spoilage.

To dress a deer in the field, you can either tie the deer's legs to a tree with cords to give you room to work or while it's lying on its side if there is no place to elevate it.

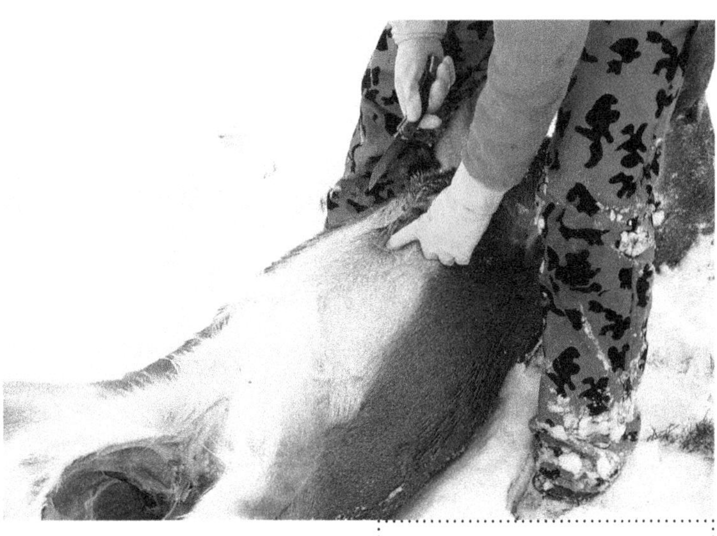

1. Field dress the animal immediately to drain the blood and dissipate the body heat. Wearing rubber gloves will protect you from any parasites or blood-borne diseases the animal may be carrying and make cleanup easier. Locate the base of the breastbone and then make a shallow cut that is long enough to insert the first two fingers of your hand, being careful not to puncture the intestines when cutting.

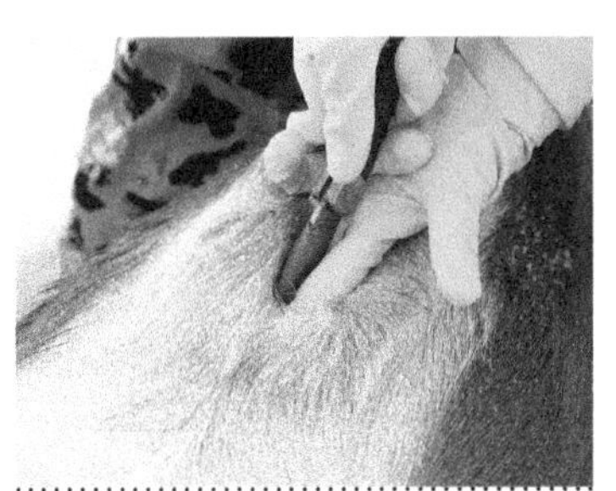

2. Form a V with the first two fingers of your hand. Hold the knife between your fingers with the cutting edge up. Cut through the abdominal wall to the pelvic area using your fingers to prevent puncturing the intestines.

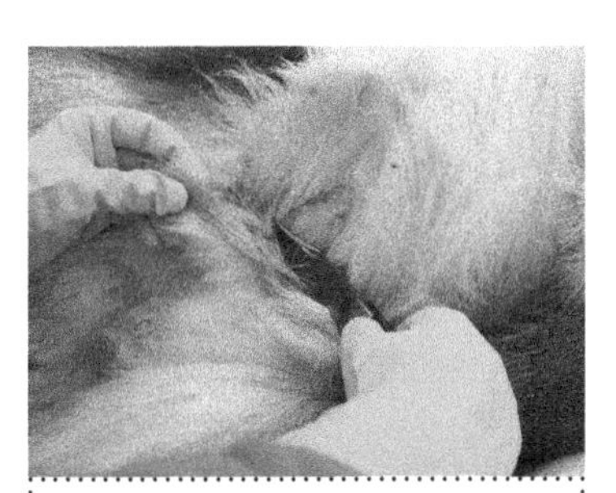

3. Separate the external reproductive organs of a buck from the abdominal wall, but do not cut them off completely. Remove the udder of a doe if it was still nursing. The milk sours rapidly and could give the meat an unpleasant flavor.

First remove the genitals if you have shot a buck, but be careful when cutting the hide in the abdomen area. If you need to drag the carcass some distance later, you will want to make as small a cut as necessary to remove the viscera. Dragging or carrying the deer will expose it to weeds, soil, and insects, which may contaminate it.

Once you have removed the genitals, cut a small opening in the hide in front of the aitchbone. If you have elevated the carcass, you can make a downward cut with the knife blade pointing outward and use the heel to slice down the inside. If your deer is lying on its side, elevate the pelvis to begin your cut down the abdomen. Then, lay it back on the ground and use your free hand to push intestines away from your knife.

Cut until you reach the rib cage, and tilt the carcass sideways to drop the viscera and drain any blood from the body cavity. Cut around the anus and pull it back through the abdominal cavity after tying it closed to prevent fecal contamination.

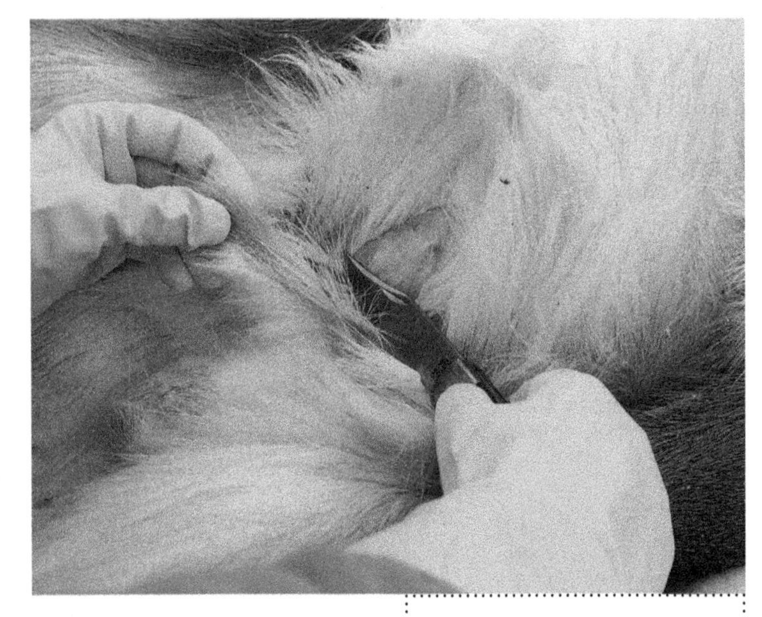

4. Straddle the animal, facing its head. If you do not plan to mount the head, cut the skin from the base of the breastbone to the jaw, with the cutting edge of the knife up. If you plan to mount the head, follow your taxidermist's instructions.

These cuts should let you pull the rest of the intestine and stomach from inside the body. You can then remove the heart, lungs, liver, esophagus, and windpipe by cutting as far forward in the chest cavity as possible. Cut the heart and liver free and save them for making sausage. Place the heart and liver in a portable cooler to keep them fresh.

Many hunters tag their deer after they have dressed it; some before. Either way, you will need to tag the deer before you load it into your vehicle. If the carcass is too large to drag or carry to your vehicle from where you dress it, you may have to split or quarter the carcass. Knowing and understanding the laws and regulations regarding the identification of and dressing of large game animals in your hunting area will eliminate potential problems if the carcass needs to be divided. Normally you will have a specified period of time in which to register your deer at a designated check station.

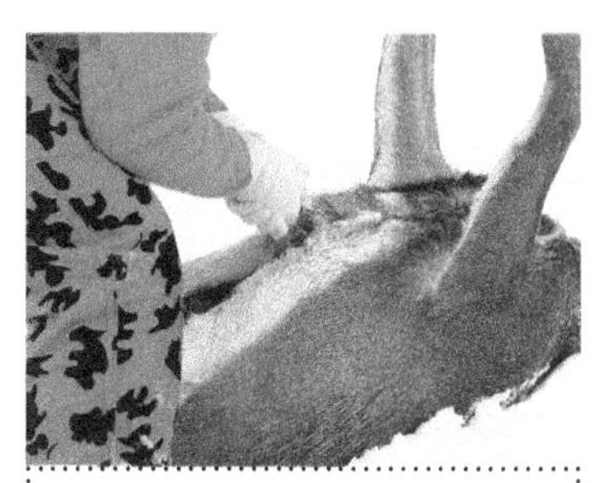

5. Cut through the center of the breastbone by bracing your elbows against your legs, with one hand supporting the other, and use your knees to provide leverage. An older animal may require using a game saw or small axe.

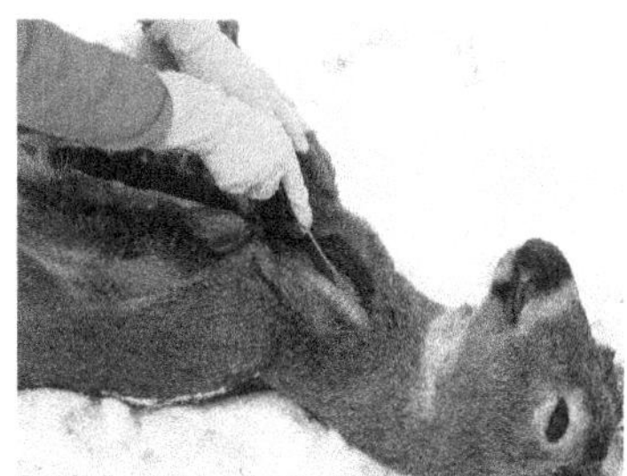

6. Free the windpipe and esophagus by cutting the connective tissue. Sever the windpipe and esophagus at the jaw. Grasp them firmly and pull down, continuing to cut where necessary, until freed to the point where the windpipe branches out into the lungs.

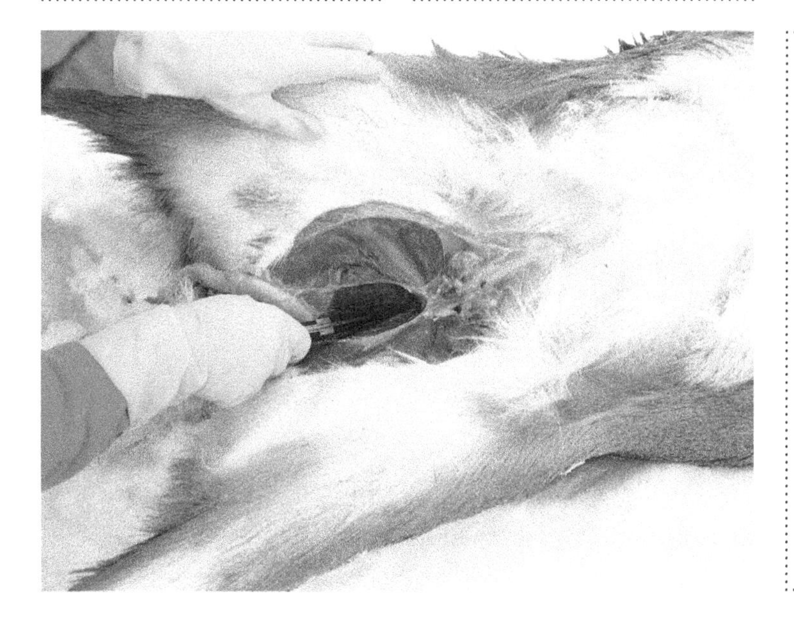

7. To free a buck's urethra, slice between the hams or split the pelvic bone on either a buck or doe. Make careful cuts around the urethra until it is freed to a point just above the anus. Be careful not to sever the urethra. Cut around the anus; on a doe, the cut should also include the vulva above the anus. Free the rectum and urethra by loosening the connective tissue with your knife. Tie off the rectum and urethra with sturdy string to prevent fecal contamination of the inside body cavity.

# CARCASS DISPOSITION

If you intend to process the carcass yourself, be sure to keep it in temperatures that do not exceed 40°F (4°C) while it is aging. You can age a deer carcass for a week before cutting it up, as this should improve its tenderness and palatability. During aging, keep the hide on to reduce moisture loss or shrinkage and to avoid discoloration of the meat.

If you don't process the carcass yourself, you may be able to make arrangements with a local meat and locker service to do it. Federal and state regulations require that wild game not be processed in conjunction with domestic animals. If you are home processing, avoid handling wild and domestic carcasses at the same time. Be aware that a wild game carcass must be dressed before it enters the processing or refrigerated areas of a licensed facility. Also, it is essential that all equipment you use that comes in contact with wild game be thoroughly cleaned and sanitized before you again use it on domestic animal or poultry carcasses.

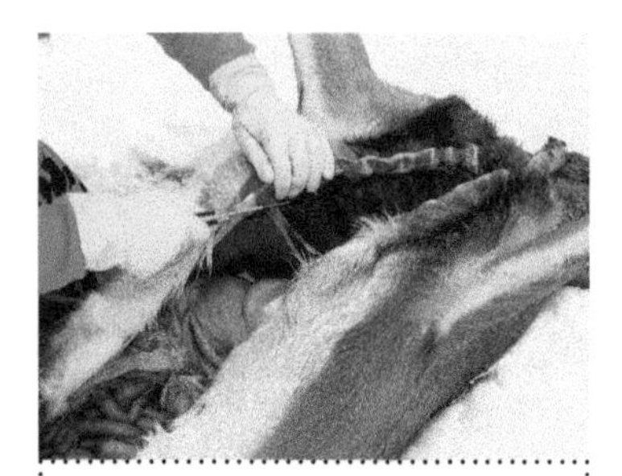

1. Hold the rib cage open on one side. Cut the diaphragm from the rib opening down to the backbone. Stay as close to the rib cage as possible; do not puncture the stomach. Repeat on the other side so that the cuts meet over the backbone.

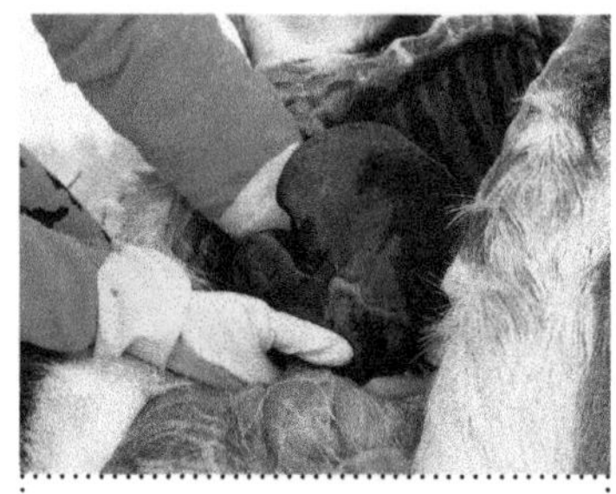

2. Cut the tubes that attach the liver and remove it. Check for spots, cysts, or scarring, which may indicate parasites or disease. If any are present, discard the liver. If the liver is clean, place into a plastic bag with the heart. Place on ice as soon as possible.

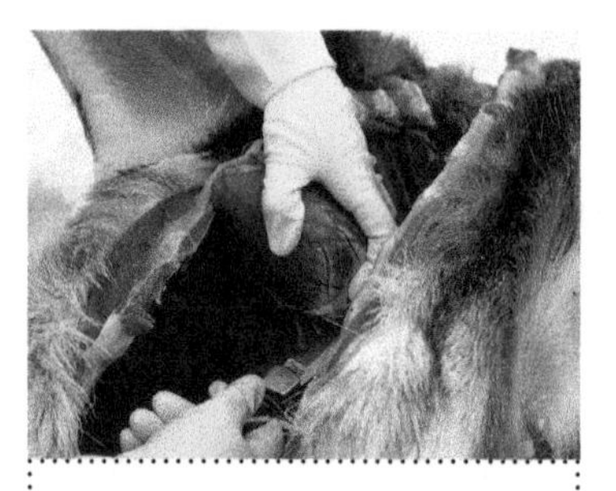

3. Remove the heart by severing the connecting blood vessels. Hold the heart upside down for a few moments to drain excess blood and then place it in a plastic bag. Some hunters find it easier to remove the viscera first, and then take the heart and liver from it.

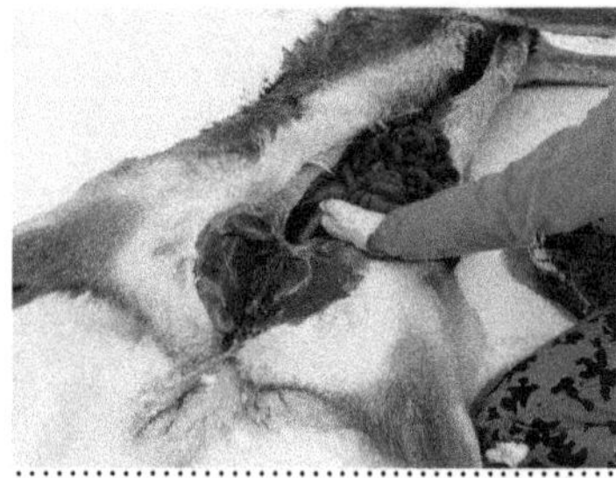

4. Pull the tied-off rectum and urethra from the pelvic bone and into the body cavity, unless you split the pelvic bone, making this unnecessary. Roll the carcass on its side so that the viscera begin to spill out the side of the body cavity.

1. Firmly grasp the windpipe and esophagus, and pull down and away from the body. If the organs do not pull away freely, the diaphragm may still be partially attached. Scoop from both ends toward the middle to finish rolling out the viscera.

## SKINNING

Elevating the carcass allows you to skin a deer easily and effectively. Before raising it, cut slits in the skin between the rear leg bone and the tendon of the hock. Insert hooks and a strong piece of wood or metal bar into the slit. This will let you raise the carcass to a level that is comfortable to work with. If it is a buck and you wish to preserve the head, take extra care when you elevate it to a height where the horns no longer touch the floor or ground to avoid breaking them.

The hocks should be spread apart to give you easier access to the abdominal area. You can make the first cuts for skinning before you elevate the carcass. First, make a complete circular cut around each hock just below the inserted hooks, and avoid cutting into the tendons. Place the blade tip on the top of the tendon and carefully slice toward the rectum. Do not cut into the hindquarter. Do the same for the other leg. From this point on, you will need very little knife work because the skin can be easily pulled and fisted from the carcass.

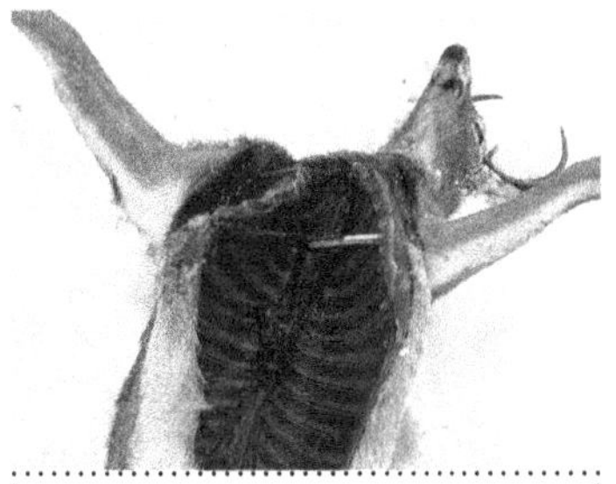

2. Sponge the cavity clean, and prop open with a stick. If the urinary tract or intestines have been severed, wash the carcass with snow or clean water. If you need to leave the carcass, drape it over brush or logs with the cavity down, or hang it from a tree to speed cooling.

3. When moving the carcass, leave the hide on to protect it from dirt and flies. An intact hide prevents surface muscles from drying too much during aging. Drag a deer with each front leg tied to an antler to keep from snagging brush, or tie a rope around the neck if antlerless. In dusty terrain, you should tie the carcass shut, then drag it on a heavy tarp to avoid damaging the muscles.

Remove the forelegs by making cuts just below the knee at the smooth joints. Then, begin pulling the hide from the rounds or rump and inside the rear legs with even tension. You may have to work the inside skin free before pulling from the top part of the anus. Use your hand or fist to remove the skin from the sides as you pull it down the back.

If you plan to mount the head, you will need to retain enough of the hide for a cape. Open the skin on the top side of the neck and behind the shoulder to make enough for the cape. Leaving too much skin available for a taxidermist is better than too little.

To remove the head, you should cut at the atlas joint so that it and the cape of skin can be removed in one piece. After the head is removed, you can split the underside of the neck and remove the remaining esophagus, windpipe, and any other part, such as the lungs and heart, if they have not already been removed. Then, brush and wash the inside body cavity with clean water to remove any hair or soil attached to it. After a thorough washing, you are ready to cut up the carcass.

## HEAD MOUNTING

If you desire to have the head mounted, you will need to care for it and the hide before taking it to a taxidermist. To help preserve it, you should liberally apply salt to the head and rub it into the skin side of the hide. Let the salt be absorbed for 24 to 48 hours before folding the skin together with the hair side out. Tie it and tag it according to the laws pertaining to your area before delivering it.

1. Skin the neck area before sawing off the head. Skinning the neck first will eliminate the chance of forcing hair into the meat with the saw.

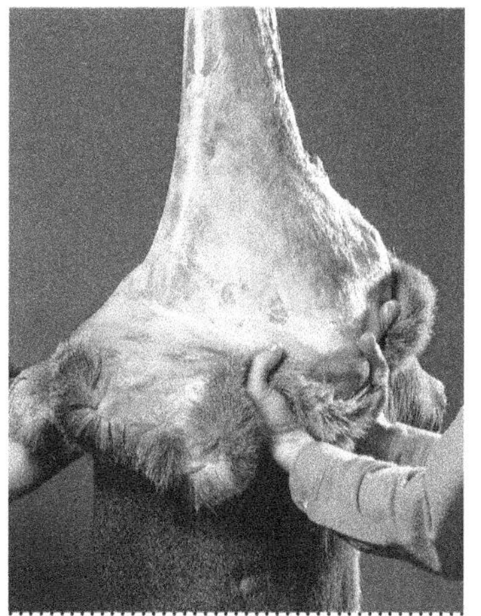

2. Skinning a deer is similar to that of a beef or pig. Start with cuts at the inner parts of the hind legs, peel the hide away, sever the tailbone, and continue peeling with your fist along the back, using your knife only when necessary, until reaching the head, which can be cut off at the atlas joint.

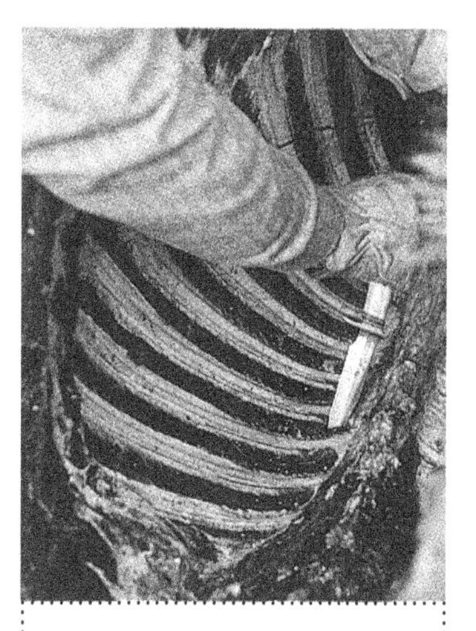

3. The evisceration process is similar to that used with deer. After the viscera are removed, cut between the third and fourth ribs, from the backbone to the tips of the ribs. Make your cuts from inside the body.

4. Split the hide along the backbone on both halves, and then peel it back several inches (18 cm) on each side of the cut to expose the spine for leaner sawing.

# CUTTING THE CARCASS

You can divide the carcass by splitting the aitchbone and sawing down the center of the backbone. Remove the neck first before you split the carcass if you plan to use the neck for pot roast or neck cuts and don't need to keep the head.

Lay the carcass on its side on a clean table, abdomen-side facing you, and begin by removing the hind legs. You can now split one side into three pieces: hindquarter, ribs, and shoulder. Make your first cut just in from and close to the hipbone. Then, separate the shoulder from the ribs and loin by cutting between the fourth and fifth ribs. The breast or flank is removed by cutting across the ribs about 3 inches (7.5 cm) from the backbone, from front to back. Then, you can separate the ribs from the loin by cutting directly behind the last rib.

To remove the rump from the leg, you can turn the aitchbone upward and make a saw cut parallel to it. You can then remove the flank with your boning knife.

The shanks, breast, and flank are generally boned and ground into burgers or mixed with pork fat for sausage. The neck slices can also be boned for ground meat and sausage. Venison rib chops, boneless tenderloins, round steaks, and rolled shoulder roasts are the most important cuts. Many of the principles used for cutting a lamb carcass can be applied for deer processing.

1. To separate the front half of the carcass from the rear half, use your saw to cut through the backbone after making your first knife cut. A quartered hide is still suitable for tanning.

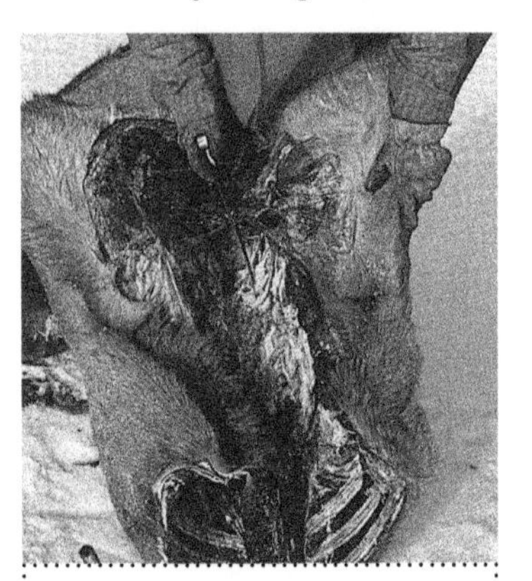

2. Begin sawing lengthwise through the backbone by propping one half against your legs. Be careful to saw down the middle of the spine and not through any of the loin.

3. Keep the back off the ground as you continue cutting. Gravity will help pull the quarters apart so that your saw doesn't bind, as it would if the half were lying on the ground.

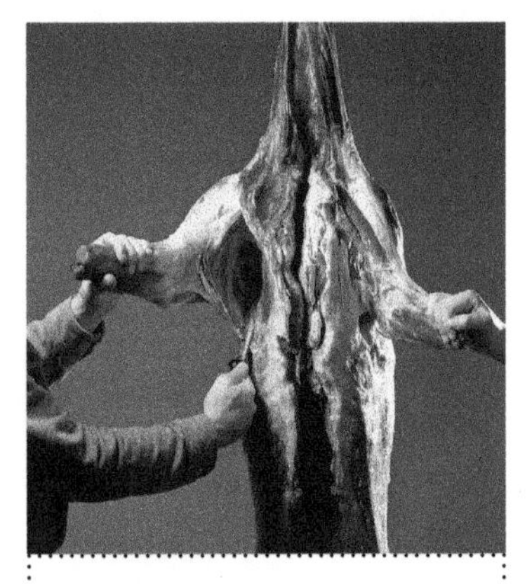

4. Begin fabrication by pushing the front leg away from the body and cut between the leg and the rib cage. Then, continue until reaching the shoulder.

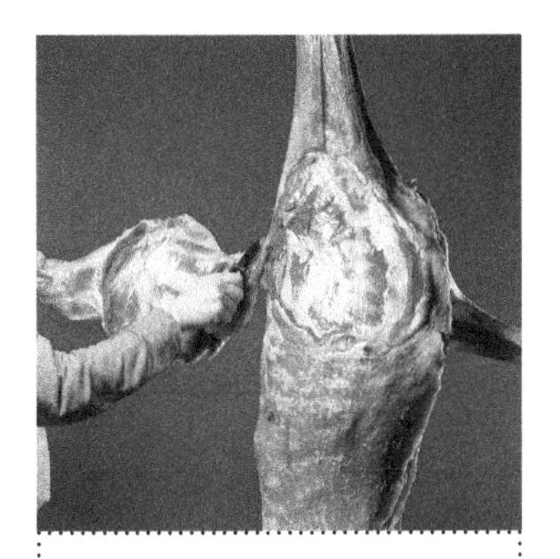

5. Remove the front leg by cutting between the shoulder blade and the back. Repeat with the other leg. Remove the layer of brisket meat over the ribs (right). Grind thin brisket for burger.

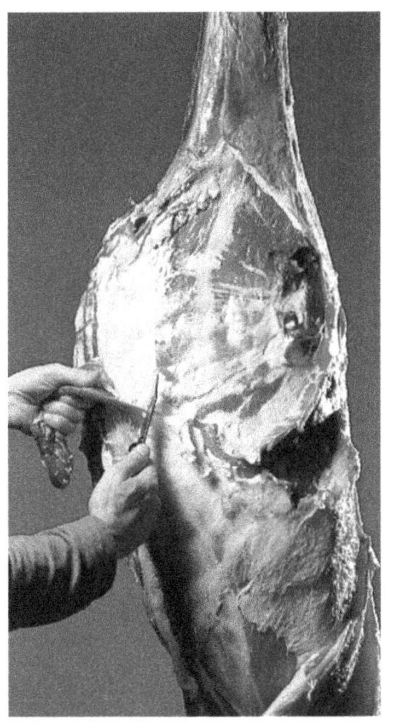

6. Cut the meat at the base of the neck to begin removing a backstrap. There are two backstraps, one on each side of the spine. They can be butterflied for steaks, cut into roasts, or sliced thinly for sautéing. The lower part, or loin, is the most tender.

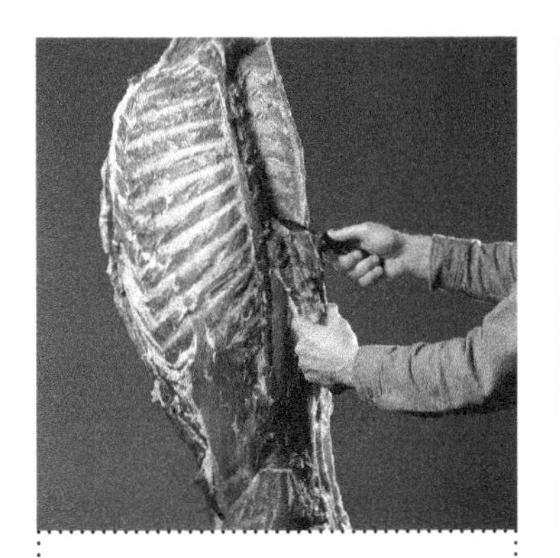

7. Make two cuts between the shoulder and rump bone along the spine and the other along the rib tops. Keep your knife close to the bones, removing as much meat as possible. Cut off this first backstrap at the rump, and then remove the backstrap on the other side of the spine.

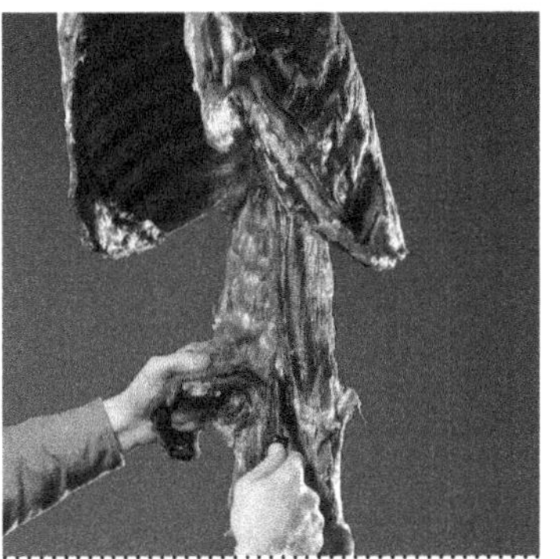

8. Cut the tenderloins from inside the body cavity after trimming the flank meat below the last rib (right). The flank meat can be ground or cut into thin strips for jerky. Many hunters remove the tenderloins before aging the carcass, to keep them from darkening and dehydrating.

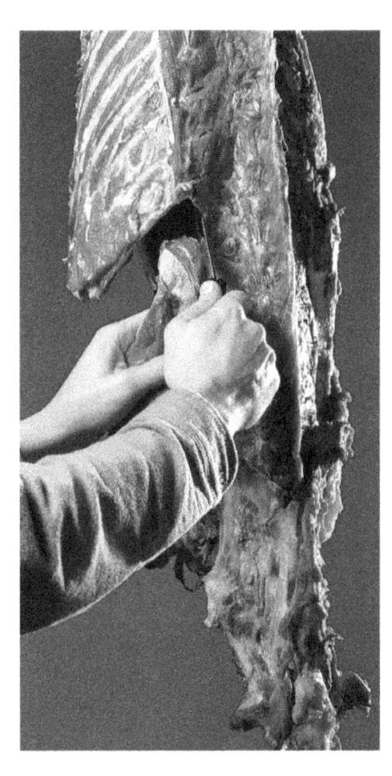

9. Begin cutting one hind leg away, exposing the ball-and-socket joint. Push the leg back to pop the joint apart, and then cut through the joint. Work your knife around the tailbone and pelvis until the leg is removed. Repeat with the other leg.

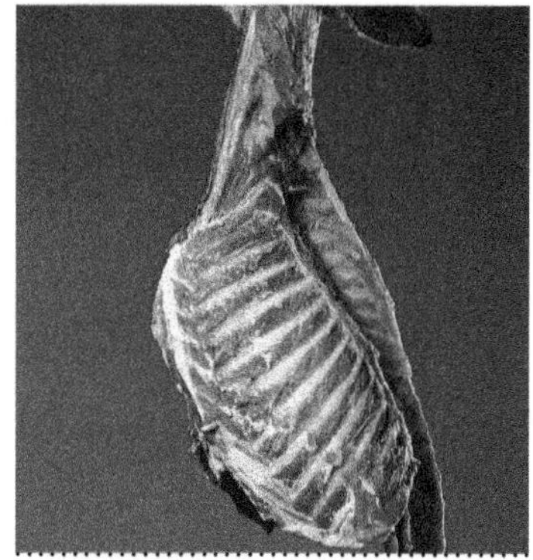

10. Remove the ribs if desired by sawing along the backbone. Cut around the base of the neck, and then twist the backbone off. Separate the neck and head. Bone the neck to grind for burgers or keep it whole for pot roasting.

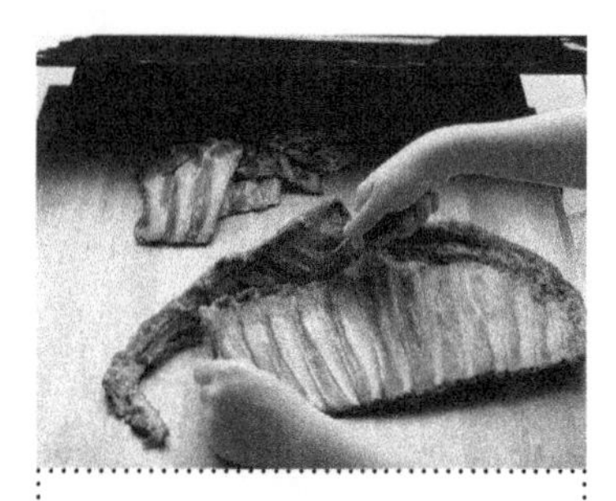

11. Trim the ribs by cutting away the ridge of meat and gristle along the bottom. If the ribs are long, saw them in half. Cut ribs into racks of three or four. If you don't want to save the ribs, you can bone the meat between them to grind for burgers or sausage.

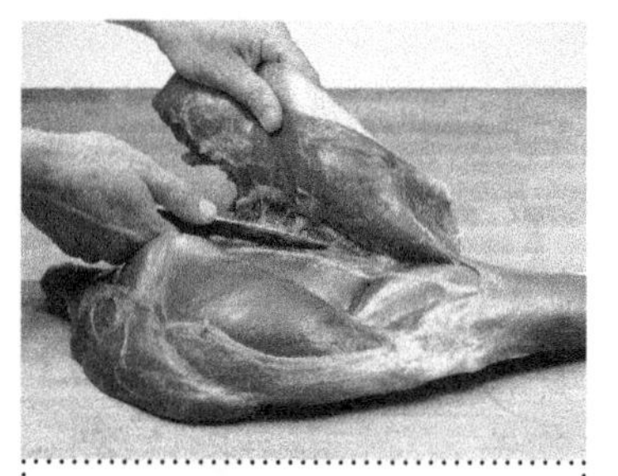

12. Cut along the back of the leg to remove the top round completely. The top round is excellent when butterflied, rolled, and tied for roasting. Or cut it into two smaller flat roasts, cubes for kabobs, or slice for sautés.

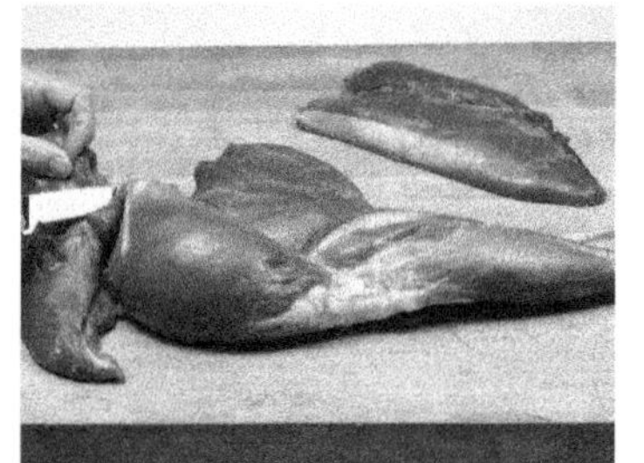

13. Remove the rump portion. Cut the rump off at the top of the hipbone after removing the silverskin and pulling the muscle groups apart with your fingers. A large rump is excellent for roasting; a small one can be cut for steaks, kabobs, or sautés.

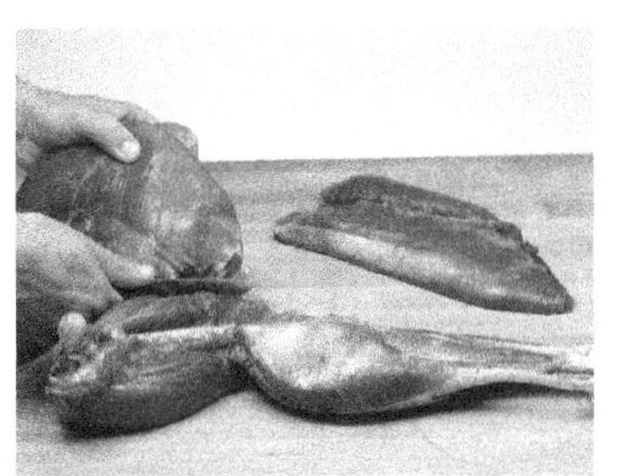

14. Cut the bottom round away from the sirloin tip after turning the leg over and separating these two muscle groups with your fingers. Next, carve the sirloin tip away from bone. Sirloin tip makes a choice roast or steaks; bottom round is good for roasting, steaks, or kabobs.

15. Large-diameter steaks can be made from a whole hind leg by cutting across all the muscle groups rather than boning as before. First, remove the rump portion, and then cut the leg into 1-inch (2.5 cm)-thick steaks. As each steak is cut, work around the bone with a fillet knife, and then slide the steak over the end of the bone. Continue steaking until you reach the shank.

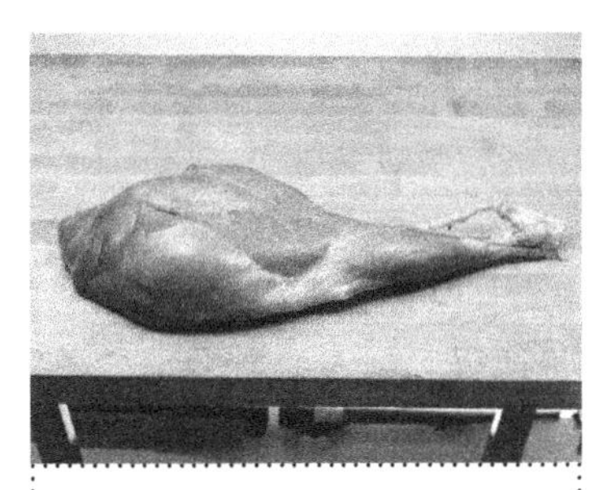

16. The hind leg consists of the sirloin tip, the top and bottom rounds, the eye of the round, a portion of the rump, and the shank. The sirloin, rounds, and rump are tender cuts for roasting or grilling; the shank is tough and best used for ground meat or soups.

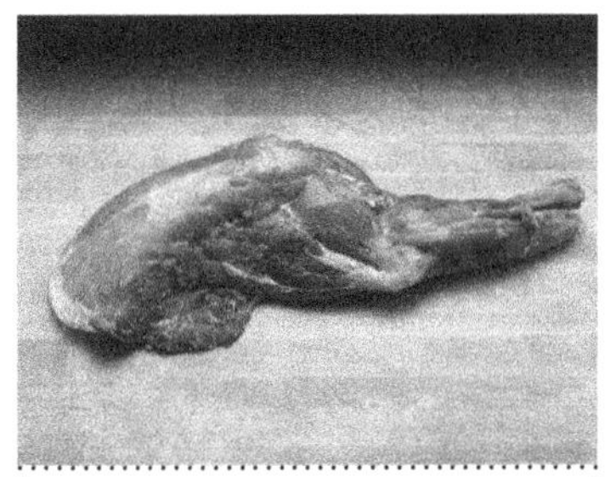

17. A front leg consists of the shoulder, arm, and shank. The meat from the front leg is less tender than that from the hind leg, and it is used for pot roasting, stews, jerky, or grinding.

# MUTILATED AREAS

Portions of the carcass may have sustained damage from a gunshot or arrow wound, depending on the season. This damaged meat may have materials imbedded in it, such as hair, metal shards, and any mixture of blood, bone chips, and fecal matter. Carefully cut out damaged tissue and dispose of it.

# MEAT VOLUME

Some estimates can be made of the amount of edible meats that can be derived from deer carcasses. Many factors will influence the weight of the animal, including its age and diet, but the percentages will remain fairly typical. A 100-pound (45.5 kg), field-dressed deer will typically be about 1.5 years of age and can dress out at up to 80 percent. This yields a carcass weight to cut of 80 pounds (36 kg). Roughly 50 percent of this, or 40 pounds (18 kg), will be edible meat, while the other half will consist of bone, fat, and mutilated areas or areas affected by shot and the resulting blood damage.

CHAPTER 8

# MEAT BYPRODUCTS AND FOOD PRESERVATION

**EARLY PEOPLES USED AS MUCH OF THE ANIMAL'S CARCASS AS POSSIBLE FOR FOOD, CLOTHING, TOOLS, WEAPONS OR BINDINGS, AND ORNAMENTS, AMONG OTHER THINGS. STOMACHS, BLADDERS, AND SKINS WERE FASHIONED INTO CONTAINERS. NOTHING WAS WASTED FOLLOWING A SUCCESSFUL HUNT FOR MEAT. TODAY'S COMMERCIAL MEAT PROCESSORS MAKE A SIMILAR USE OF THE ENTIRE CARCASS. MODERN SOCIETY STILL UTILIZES THE NONMUSCLE PORTIONS AND TRANSFORMS THEM INTO MANY PRODUCTS, INCLUDING PHARMACEUTICAL USES.**

All parts of the animal that are not included in the carcass are called byproducts. These include such parts as skin, bones, hair, teeth, feathers, claws, fat, brains, and nonconnective tissues and tendons. Skins may be sold or tanned for leather, feathers can be used for ornaments, and fats can be used in sausage making or rendered into lard.

Edible byproducts, such as livers, hearts, tongues, testicles, kidneys, oxtails, stomachs (tripe), and intestines, may find uses in your meals in one form or another. These are sometimes referred to as variety meats.

Several, such as liver, kidney, and heart, are high in protein content and highly nutritious. These variety meats are generally more perishable than other meats and should be frozen or cooked soon after harvest or purchase.

The paragraphs that follow detail these byproducts, how they can be cooked, and for what purpose.

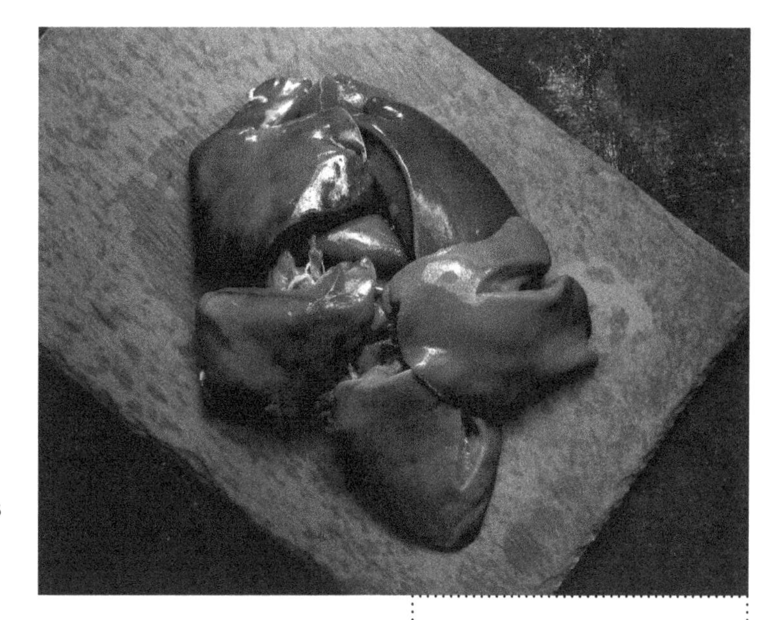

*Liver:* After removing all connective tissue surrounding the liver, it can be thinly sliced and cooked in a variety of methods. These include frying, broiling, sautéing, and braising. You can grind or chop liver and use it as an additive to sausages, loaves, spreads, and other dishes.

You may develop a taste for a particular byproduct part, as most can be included in specialized dishes or can be ground into meat for making sausage.

*Heart:* The heart from different animals can be used. For fowl and small animals, such as squirrels or rabbits, they can be cooked with moist heat or ground and used in sausages. The heart is generally less tender than liver, although it has an excellent flavor. Large animal hearts can be sliced open for inspection and then may be filled with a dressing, stitched shut with cooking thread, and then roasted with moist heat, like a turkey.

*Tongue:* Tongues from large animals, such as cattle or pigs, can make cold sandwich meats after being braised and thinly sliced. You can remove the tough outer membrane of the tongue by blanching, followed by moist heat cooking for an extended period. Once this membrane is trimmed, the rest can be cooled and sliced.

*Kidneys:* Lamb and veal kidneys may be broiled and skewered and are more tender than beef kidneys. They can be included in meat casseroles, stews, and other dishes.

*Oxtail:* The oxtail is the upper portion of the tail that attaches near the lateral end of the spine. It is often used as soup stock. The tail sections can be browned first, if desired, then simmered until the meat is tender and separates from the bone. Remove the bone parts and use the meat as soup ingredients. Oxtail has a rich, meaty flavor and texture, and it can also be used in stews.

*Tripe:* The edible portion of stomach tissues is called tripe. Tripe comes from the first and second stomachs of cattle and the stomach of pigs. To use tripe, you must cut open the stomach to remove all contents. Then, it must be thoroughly washed before quickly chilling it. Remove the inner surface membrane. It will have a firmer texture than other variety meats and is less tender. You can use it in various meat dishes, such as kidneys, or added to soups. It is best cooked with moist heat and may be served with sauces and dressings.

*Intestines*: If the intestines are not used as casings for sausage making, they can be utilized for specialty dishes. The intestines, especially those from pigs, can be cleaned and thoroughly washed and cooked with sauces. These are sometimes referred to as chitterlings. They can also be cut into small pieces, breaded with raw eggs and bread crumbs, and deep fried.

*Blood*: Blood contains approximately 17 percent protein and can be used in sausage making; blood sausage makes the best use. The U.S. Department of Agriculture regulations do not allow any blood that comes in contact with the surface of the body of an animal or is contaminated in any other manner to be used for food purposes. Only blood from inspected animals can legally be used for meat food products that are sold to the public. Some societies, such as the Masai tribes, will mix blood with milk as part of their diet, but blood is generally discarded in home butchering. Cattle have the largest amount of blood, and a 1,200-pound (544 kg) animal will typically yield about 46 pounds (21 kg) of blood; a pig will yield about 7 percent of its live body weight, and sheep will yield about 3 percent of its live body weight.

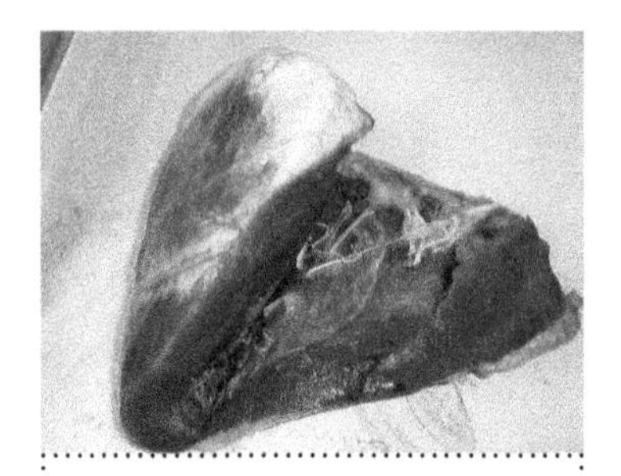

The heart is less tender than the liver, but it has excellent flavor. Small animal hearts can be ground as an additive for sausages. After the heart is removed from the body, you should slice it open and wash out any remaining blood.

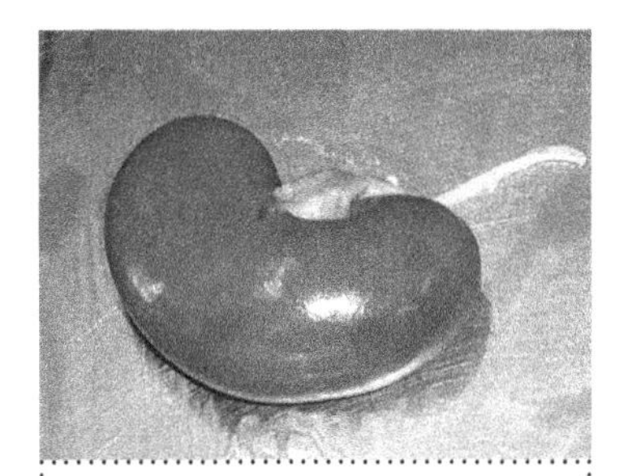

The kidneys can be used in casseroles and stews, or they may be broiled and skewered. While in the carcass, they are often surrounded by fat, which should be peeled away before use. Lamb and veal kidneys are more tender than beef kidneys.

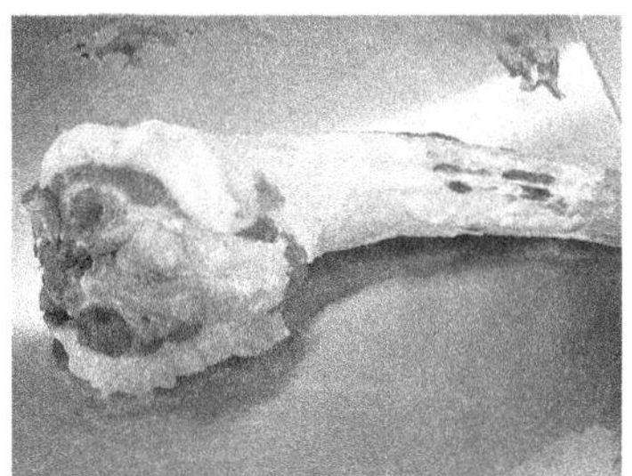

The oxtail is often used in soup stock. It is the top portion of the tail that attaches to the posterior end of the spine. It can contribute a rich, meaty flavor.

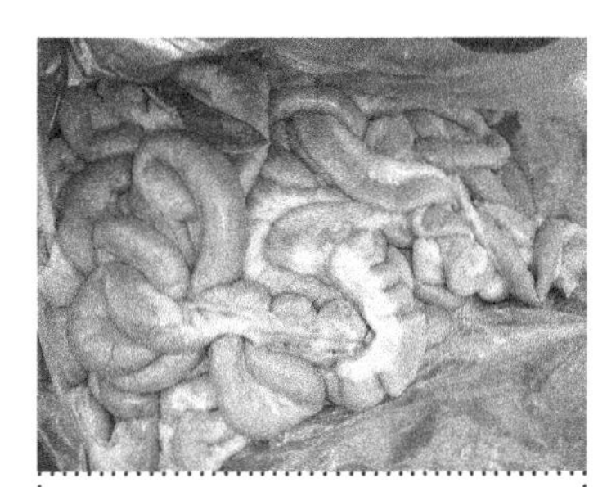

The intestines can be used for casing in sausage making. Pig intestines are the most commonly used after being thoroughly washed and brined. They can also be cooked and cut into small pieces, breaded with raw egg and crumbs, and then deep fried.

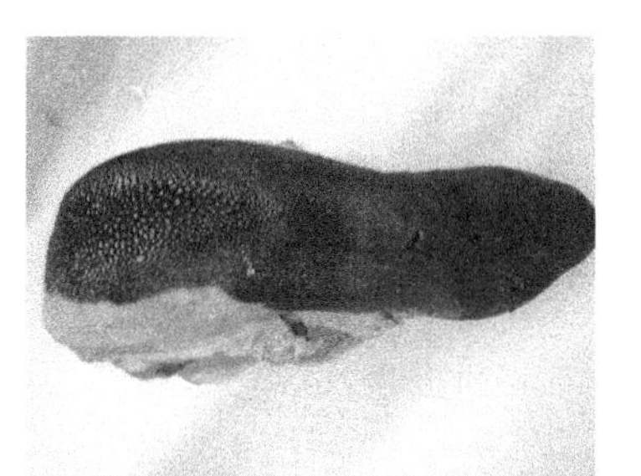

Tongues can be thinly sliced after braising for cold sandwich meats. You can remove the tough outer membrane by blanching (short exposure to boiling water) prior to long-term, moist-heat cooking. Sweetbreads include the beef and veal thymus and pork pancreas glands and are similar to tongues in that they need to be precooked to remove the outer membranes and sliced thinly before dipping in batter or flour and deep fried.

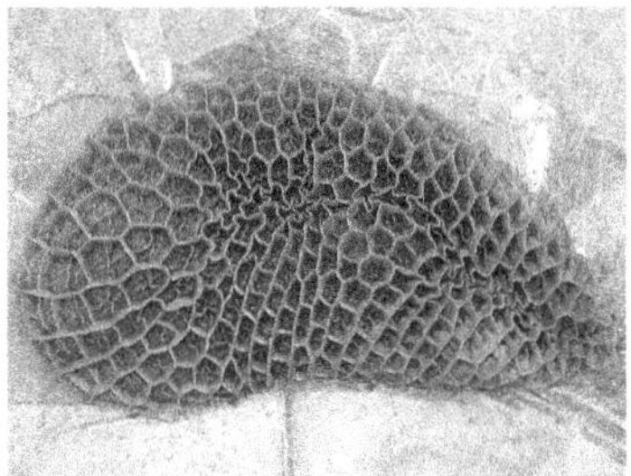

The honeycomb edible portion of a beef animal's stomach is called tripe. Before use, it must be opened, the contents thoroughly cleaned out, and then cooled. After the inner surface membrane is removed, it can be sliced and cooked with moist heat, or used in soups.

Blood contains about 17 percent protein and can be used for making blood sausage. Pork or beef blood is most often used. It is added with fillers such as meat, fat, barley, oatmeal, and bread until it is thick enough to congeal when cooked. Any blood that comes in contact with the surface of the carcass or is otherwise contaminated should not be used.

*Fats*: Animal fats have had many uses over the course of human history. They have served as a food energy source, been rubbed into animal hides to make tepee leather supple and waterproof, and, in the case of bear grease during the nineteenth century, provided a sheen and unique aroma to human hair. Fats also have been used for making soaps and providing fuel for oil lamps in the days prior to kerosene or gasoline.

Animal and plant fats differ mainly in melting point and saturation. Fats and oils contain both saturated and unsaturated components. Saturated fats are firmer and have a higher melting point than the softer, unsaturated fats.

Cooking and table fats available for use range from liquid oils, derived mainly from plants, to solid fats, which come from animals and other sources.

Lard is the fat most often used in home cooking and is rendered from the clear and edible tissues of pigs. Recent decades have witnessed a decreased in use of lard production due to health concerns and competition from vegetable fats. However, lard provides a source of energy, and linoleic fatty acid is an essential component of the human diet.

Lard has a melting point that is near body temperature, making it easily digested. This low point allows a cook or baker to use it in a variety of ways, such as cooking fat, shortening, a flavor ingredient, and a source of nutrition.

Rendering is the process of extracting fat from the tissues using heat. The raw fat and meat is either cooked or heated to turn the fat into a liquid. The melted fat is then drawn off. This process increases the shelf life of the fat by killing the microorganisms that were present and removing most of the moisture.

You can render pork fat at home whether it comes from butchering a pig or is purchased at a local market. If you are using one of your pigs, remove the raw fat from the skin to obtain better quality.

Begin by chopping the fat into fine pieces. For each pound (455 g) of fat, allow 1/2 cup (120 ml) of water. Place the fat and water in a cooking vessel and heat to boiling, but do not exceed 240°F (116°C). Stir as it warms to avoid scorching. As it boils, the steam will remove extraneous odors. Boiling will not occur until the fat liquefies. Allow the fat to cook until the solid material reaches a golden or amber color, then drain into storage containers.

Use several thicknesses of cheesecloth or similar material that can be placed over clean, dry, nonmetallic storage containers suitable for use with hot liquids. Slowly pour the hot fat into the cloths until the containers are fitted to a desired level. Cool the lard rapidly to produce a firm, smooth-textured product. As the lard cools, stir it occasionally when it reaches the creamy stage to reduce the oils separating out and avoid developing a grainy texture. Store at temperatures of 40°F (4°C) or lower. Lard may be frozen but should be packaged in airtight containers and used within six months. This will reduce changes in flavor or aroma due to oxidation.

If you choose to render fat in your home, you need to use caution during the entire process. You also need to avoid spills because hot grease can cause severe burns to exposed skin. If the fat is spilled on clothing, it will cling to it and can also cause deep, severe burns. Never allow children anywhere near your processing area or as the containers are cooling.

## FOOD PRESERVATION OPTIONS

Whether butchering domestic animals or hunting wild game for meat, it is likely you will have more meat available than you can eat at one meal. The rest will have to be preserved to keep it usable for later. Different preservation methods serve different purposes. You should decide on a plan of distribution of the meat into one or several of these methods before you proceed with processing. Do you want to freeze all or part of the carcass that is not used immediately? Do you want to make some into sausages? Develop a plan before you begin and you will eliminate waste.

Food preservation, in relation to meat products, is the process of handling it in a way that stops or retards the growth of microorganisms, making it safe for long-term consumption. There are many forms of food preservation, including freezing, canning, drying, salting, and pickling to name the major processes used in homes. Large commercial applications employ methods such as vacuum packaging, irradiation, sugaring, and using lye, modified atmosphere, and high pressure, but those will not be discussed here.

Freezing is one of the most commonly used preservative methods and has several advantages. It is a fast and simple way to stop microbial growth. The nutritional value of the meat does not deteriorate through freezing, although some texture and quality can be affected by long-term storage, resulting in freezer burn.

Canning involves cooking food to a boiling point for a specified time as a form of sterilization. This is done while sealed cans or jars are submerged in boiling water or placed in a pressure cooker. The advantages of canning include a ready-to-heat meal after the can or jar is opened, it can serve as a backup in case of freezer malfunction, and it allows useful preservation when insufficient freezer space is available.

Pickling involves the use of brines or vinegars to preserve meat products through fermentation. It is mostly used for fruits and vegetables, but it can be used for animal and fish products, such as pig's feet, pork hocks, corned beef, herring, northern pike, or other large game fish.

Drying is perhaps the oldest method of food preservation and involves dehydration of the meat. The removal of water from the meat significantly reduces the water activity to prevent, inhibit, or delay bacterial growth. Reducing the amount of water in meat also reduces the total weight, making it easier to transport.

Salting is used to cure meats by drawing moisture out of the tissue through a process of osmosis. Salt or sugar can be used separately or in a combination. Salted fish or meat was a staple in the diets of many early settlers who were on the move or lacked other methods to preserve food.

Pickling is the use of a brine, vinegar, or any spicy edible solution that is used to inhibit microbial action. Pickling can involve two different forms: chemical pickling, which uses a brine solution, or fermentation pickling, such as making sauerkraut. The edible liquid used in chemical pickling typically includes agents such as a high salt brine, vinegar, alcohol, and vegetable oils, particularly olive oil. The purpose is to saturate the food being preserved with the agent. This may be enhanced in some cases with heating or boiling. Common foods that are chemically pickled include corned beef, peppers, herring, eggs, and cucumbers. Fermentation pickling is generally not used with meat but can be used with foods that offer a good complement to meat during a meal, such as sauerkraut. Fermentation pickling is assisted by the food being preserved as it produces lactic acid.

Until the middle of the last century, jugging was a popular method of preserving meat. This method involves the process of stewing meat in an earthenware jug or casserole. The meat is cut into pieces, placed in a tightly sealed jug with brine or gravy, and stewed. Sometimes red wine is added to the cooking liquid.

## FREEZER EFFECTS ON MEAT

Freezing has a physical effect on meat, but it remains one of the best preservation methods available for long-term storage while not destroying vitamins or the meat's nutritional value. Freezing meat almost completely inactivates the enzymes and inhibits the growth of spoilage organisms.

When processing a carcass, it is important to remember that meat temperatures must be brought down to 40°F (4°C) within 16 hours to prevent growth of spoilage microorganisms that lie deep within the carcass tissues or in the centers of containers of warm meat. If the meat has not been cooled but is going directly to a freezer, it must reach a temperature of 0°F (–18°C) within 72 hours to prevent the growth of putrifying bacteria.

You should make cuts to be frozen into smaller, individual sizes that are ready for cooking rather than to freeze them as large portions that need to be further deconstructed once they are thawed. Smaller cuts freeze more quickly and evenly than very large pieces or chunks. It is better to minimize the number of times the meat needs to be handled and exposed to surfaces after it is thawed and used.

## CANNING

Canning is the second most commonly used preservation method for long-term storage of meat. Canned meats are generally of two types: sterilized and pasteurized. Sterilized meat products do not need refrigeration and can sit on shelves for extended periods as long as the container remains intact. Pasteurized products require refrigeration to inhibit spoilage.

When canning meat for home use, you should use the appropriate procedures to ensure quality and safe storage. Canned meats are preserved by hermetically sealing the container, which prevents air from escaping or entering it. By applying heat to the sealed meats, you destroy the microorganisms that are capable of producing spoilage. Using proper sanitation during the breakdown of the carcass will help minimize the number of organisms originally present at canning time.

Canning involves a time-temperature relationship in destroying most microorganisms. A specific internal temperature must be reached and held for a minimum amount of time to destroy the microorganisms present. This method is most often applied to destroy the spores that can lead to botulism. These times and temperatures are at the high end of any other methods. A safe cook, which is considered to be one that destroys the botulism organisms, requires a minimum of three minutes at 250°F (121°C). Achieving this sterilizing temperature will require the use of a pressure cooker. These typically operate under pressure of 12 to 15 pounds per square inch (844 to 1,055 g/cm$^2$). Pressure changes the boiling point of water and allows it to rise above the normal boiling point of 212°F (100°C).

There are many advantages to using the sterilization method of canning meat products. The best is a long storage life over a wide temperature range. Canned meats that are properly done can last several years and still be edible, although some flavor deterioration may occur. Providing the container has remained intact and the seal or exterior has not been damaged, canning can provide an effective preservation method when refrigeration storage or electrical power disruption occurs. Some popular products that can be canned include roast beef, beef stews, canned beef, potted-meat products, and pickled pig's feet.

Glass jars are used for canning meats and vegetables. Inspect all jars for rim cracks or chips, and discard any jar when they are found because they will not create a good, safe seal. Also, inspect each rubber ring or metal lid with gaskets to be used and discard any that are defective.

The two most important aspects of canning are providing sufficient heat and creating a perfect seal of the container. Only the best and freshest meats should be used because canning only preserves the meat; it does not improve the quality of the meat used.

Glass jars are typically used in a method called the hot pack. This involves packing the meat into the jars and processing the jars in boiling water or steam. The advantages of this method are that the jars are completely sealed and the meat has no further exposure to outside influences or organisms.

Glass canning jars come in several sizes but are most generally found in pints (475 ml) or quarts (950 ml). Covers or lids that can be firmly tightened before being placed in water are needed. Rubber rings and metal lids with a sealing gasket attached to it are two popular options. New rings and seal lids must be used each year; discard used lids or rings.

Many models of pressure cookers are commercially available, such as stovetop and electric models, and they are usually made of aluminum or enameled steel. Whichever model you choose, the same principles apply. It should be substantially constructed and should have a pressure indicator, a safety valve, and a petcock or vent.

Begin by thoroughly washing each jar in hot soapy water and rinse it in clear, hot water. The jars can air dry by placing them on clean towels. Inspect the jars and test the cooker before you begin. Examine the jars and lids for nicks or cracks. If they appear intact, fit a new ring to each jar, partly fill with hot water, and adjust the lid and seal. Invert each jar and watch for leakage or small bubbles rising through the water as it cools. An imperfect seal means you should discard the jar or the lid, depending from where the bubbles originate. You can also test the rubber rings by doubling them over. If any crack, discard them.

Fill the jar half to three-quarter full with small raw cubed-sized pieces of meat. Do not tightly pack down the meat into the jar. Fill each jar with clean water to within 1 inch (2.5 cm) of the rim. Place the jars on the rack inside the cooker. The water level should reach the bottom of the rack, which keeps the glass off the chamber base and allows the water and steam to completely surround the jars. Place the jars so they do not touch one another. The lid or cover should be adjusted carefully and fastened tightly so that no steam can escape. The petcock should stay open until steam has poured out steadily for 10 minutes or more. Then, close it to allow the pressure to rise to the level directed in the owner's manual, usually 10 pounds (4.5 kg).

To begin the processing, place a small amount of water in the bottom of the cooker. Add the meat-filled jars to the cooker and clamp on the air-tight lid. The cooker is then set over heat or heat is applied electrically.

The pressure raises the temperature higher than that used in ordinary cooking, and the food cooks more quickly. A gauge on the lid shows the number of pounds or grams of pressure, indicating the temperature. A safety valve releases pressure after cooking is completed. It will also release excess pressure. The petcock provides an outlet for steam and air.

When the appropriate pressure is reached, you should adjust the heat to keep the same pressure without variation. For meats, process for three minutes at 250°F (121°C).

When the processing time is completed, the cooker should be taken off the heat and left alone until the pressure goes down to zero. Then, open the petcock to release the remaining steam. Liquid may be lost from the jars if the pressure varies during processing or if the steam is released too quickly. Jars should not be reopened and refilled under any circumstances unless being immediately used. Do not let the cooker sit unopened for any length of time after the steam is down.

Pressure cookers are used to create a higher cooking temperature than is possible under normal cooking conditions. Water heated under pressure increases the temperature quickly. Be sure the gauge is accurate, the handles are securely fastened before heating the water, and the petcock functions appropriately to ensure safe use of a pressure cooker.

This may create a vacuum, which will make it difficult to open the lid. If this happens, reheat the cooker for a few minutes until it is loose.

Take the jars out of the cooker and hand-tighten their lids if necessary. Place the jars on a rack or towel to cool, but keep them away from drafts. Some canners turn the jars upside down as they cool to check for any leaks or bubbles, which indicate a poor seal.

After drying, label the jars with the canning date. After ten days, recheck the jars. Immediately discard any that exhibit cloudiness or signs of spoilage. Do not eat their contents under any circumstances.

# MEAT COOKING METHODS

The cooking method you use depends on the kind and quality of the meat to be cooked. Only tender cuts of meat can be cooked by dry heat. Less tender cuts require moist heat and long, slow cooking. The kind of cooking methods include the following:

*Baking*: To cook in an oven or oven-type appliance. Covered or uncovered containers may be used.

*Barbequing*: To roast slowly on a spit or rack over coals or under a gas broiler flame or electric broiler unit, usually basting with a highly seasoned sauce. The term also is commonly applied to foods cooked in or served with barbeque sauce.

*Boiling*: To cook in water or mostly water, at boiling temperature (212°F [100°C] at sea level). Bubbles rise continually and break on the surface.

*Braising*: Braising is cooking by moist heat. It is used for the less tender cuts, which require long, slow cooking in the presence of moisture to bring out the full flavor and make them tender. Many pork cuts are cooked by braising rather than broiling or pan-broiling because pork requires thorough cooking. Brown the meat in a small amount of fat, then cover tightly and cook slowly in juices from the meat or in added liquid, such as water, milk, or cream. Add only a small amount of liquid occasionally and do not let boil but keep at a simmering temperature. Pork chops and pot roasts can be cooked by braising.

*Broiling*: Broiling is cooking by direct heat and may be done over hot coals or under a flame or an electric unit. This method may be used for tender cuts that have adequate amounts of fat. Veal and pork should not be broiled since they are too low in fat.

*Caramelizing*: To heat sugar or food containing sugar until a brown color and characteristic flavor develop.

*Creaming*: To work a food or a combination of foods until soft and creamy, using a spoon, wooden paddle, or other utensil.

*Fricasseeing*: To braise individual serving pieces of meat, poultry, or game in a little liquid such as water, broth, or sauce.

*Frying and sautéing*: Some meats such as chops and cutlets may be crumbed and fried in deep fat or oil. Ham, liver, and some other meats can be sautéed in a small amount of oil or fat at low temperatures after the first searing.

*Marinating*: To let foods stand in a liquid (usually a mixture of oil with vinegar or lemon juice) to add flavor or to make them more tender.

*Pan-broiling*: To pan-broil, place the meat in a sizzling skillet or pan, and brown on both sides. Reduce heat, pour off fat as it accumulates, and cook until done, while occasionally turning it. Pork is generally not pan-broiled

*Parboil*: To boil until partly cooked.

*Pot roasting*: To cook large pieces of meat by braising.

*Roasting*: To roast meat, place it on the rack in a roasting pan, fat side up, and cook in a slow oven, uncovered and without water, until cooked as desired. The large tender cuts of meat are cooked by this method.

*Scalding*: To heat liquid to just below the boiling point.

*Simmering*: To cook in liquid just below the boiling point, at temperatures of 185°F to 210°F (85°C to 99°C). Bubbles form slowly and break below the surface.

*Stewing*: To boil or simmer in a small amount of liquid. Cut the meat into cubes and brown on all sides in hot fat, if desired. Cover with boiling water and cook at simmering temperatures in a covered kettle until meat is tender. Less tender cuts containing much connective tissue are best cooked by stewing, which softens both tissue and fiber. The best cuts for stews are those containing both fats and lean and some bone. The shank is the most economical of all cuts for this purpose. Other cuts used are the neck, plate, flank, heel of the round, and short ribs. The brisket and rump are sometimes used.

# WRAPPING

Rancidity is the bad taste or smell derived from fats or oils that have spoiled. It develops differently in animal carcasses, depending on their fats' ability to absorb oxygen from the air. Rancidity can affect the taste, odor, and palatability of the fat and adjoining tissue. Different animal species produce different fats. Pork fat is high in unsaturated fatty acids, which have the ability to absorb oxygen resulting in a shorter storage life. Beef and lamb have a higher proportion of saturated fatty acids and are less susceptible to oxygen absorption and generally have a longer storage life.

You can reduce oxidation effects by eliminating air exposure to the meat. One good way for home use is to properly apply a wrapping material that is airtight and moistureproof.

Loss of meat moisture is referred to as shrinkage or dehydration. The loss of moisture from the frozen surface of the meat is called freezer burn. Freezer burn results from surface moisture loss due to using an unsuitable grade of wrapping paper, holes in the paper, or improper wrapping. Severe dehydration results in lower quality cuts, and the increase in oxygen exposure to fats can make them rancid.

To avoid freezer burn and reduce oxygen effects, you should use a good grade of meat wrapping paper that is moistureproof and use proper wrapping and handling procedures that eliminate tears or cuts in the paper. There are many American companies that supply meat wrapping paper, often referred to as butcher paper.

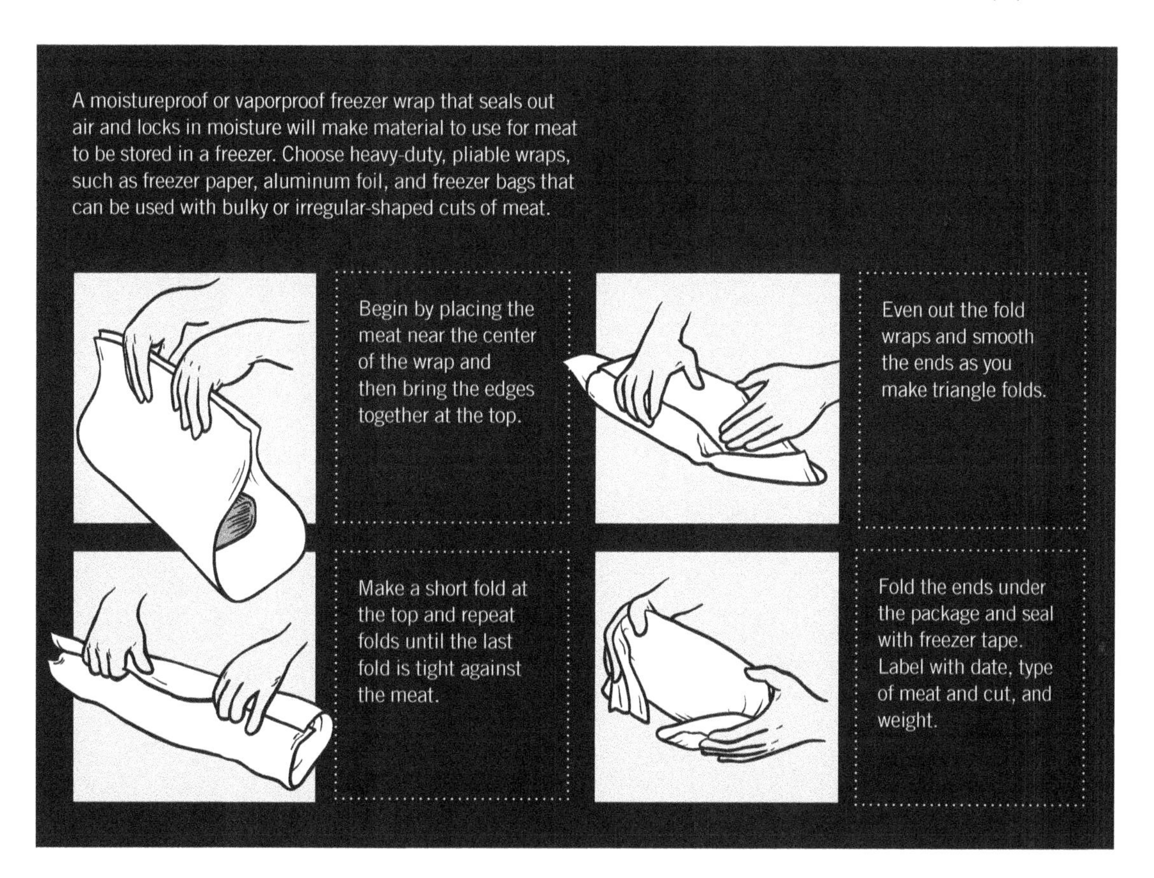

A moistureproof or vaporproof freezer wrap that seals out air and locks in moisture will make material to use for meat to be stored in a freezer. Choose heavy-duty, pliable wraps, such as freezer paper, aluminum foil, and freezer bags that can be used with bulky or irregular-shaped cuts of meat.
Begin by placing the meat near the center of the wrap and then bring the edges together at the top.
Even out the fold wraps and smooth the ends as you make triangle folds.
Make a short fold at the top and repeat folds until the last fold is tight against the meat.
Fold the ends under the package and seal with freezer tape. Label with date, type of meat and cut, and weight.

CHAPTER 9

# MEAT CURING AND SMOKING

**CURING AND SMOKING ARE TWO OF THE MOST PREVALENT MEAT-PRESERVATION METHODS USED TODAY, BOTH COMMERCIALLY AND AT HOME. BOTH TECHNIQUES INVOLVE DRYING, WHICH HELPS INHIBIT BACTERIAL GROWTH. SINCE MOISTURE IS REQUIRED FOR BACTERIA TO FUNCTION AND MULTIPLY, REMOVING IT RETARDS THAT GROWTH AND MAKES PRESERVED MEAT SAFE TO CONSUME FOR A LONGER PERIOD OF TIME.**

Curing and smoking are not only good ways to extend the shelf life of meat, they create delicious, unique flavors.

You can find charcuturie boards on the menu of restaurants around the globe.

If you're only smoking meat, you may well be smoking it with the intent to consume it immediately—or smoking it with the intent of finishing it on the grill or in the oven. If this is the case, there are fewer safety concerns. Still, you'll find curing and smoking often go hand in hand. Both can help extend the shelf life and long-term use of your meat products. Besides creating a more varied product, smoking and curing together provide unique flavors and colors for various cuts of meat. Sausages, patties, roasts, loin cuts, rib cuts, bacon, and other cuts can be infused with layers of flavor by smoking and curing.

Using a smoking or curing process may depend on your tastes and preferences, the space you have available, and perhaps your expertise. Taste does need not be sacrificed when using safe procedures. This chapter will guide you through the processes needed to make preserved meat products that are both safe to consume and tasty.

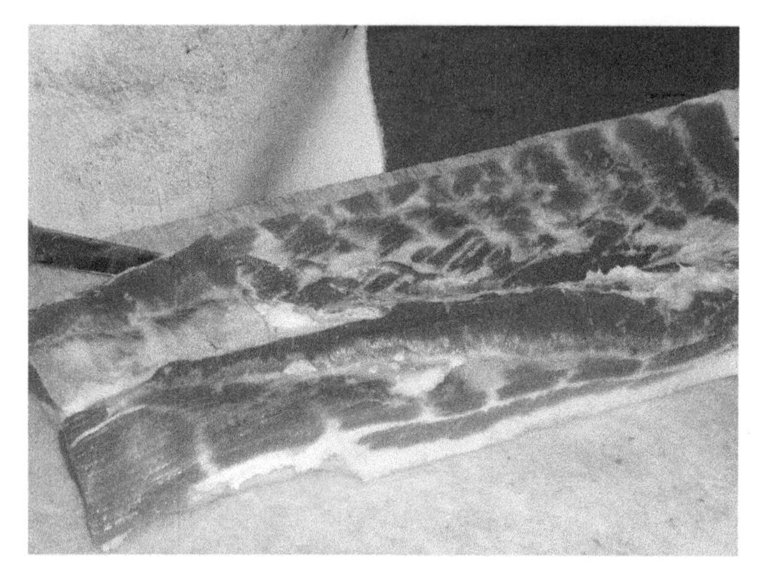

A quality meat cut makes the best end product. A pork belly, whether whole or in part, can be smoked and cured before it is cut into strips for frying or broiling.

# THE BASICS OF CURING

Curing is one of the oldest food preservation methods, and it is still one of the most useful preservation methods today. Through much of human civilization, salt has been the primary preservative used to avoid spoilage and create a longer shelf life for meat and today's meat preservation methods still routinely use salt along with nitrite—both of which help prevent the growth of microorganisms that cause botulism, a lethal form of food poisoning.

Salt is the essential ingredient in any successful curing process. Curing is generally done using either a liquid marinade into which the meat is submerged or a dry rub that is used to coat the meat's exterior. Typically, the meat must sit for a certain length of time before drying for the cure to work through the exterior surface and into the interior, drawing moisture and blood from the muscle cells while entering the cells by osmosis. The salt slowly moves through the cell membranes, replacing moisture and inducing partial drying. Salt is central to all curing mixtures because it is hygroscopic, which means it absorbs moisture from its surrounding environment, thus providing its preservative action.

Generally speaking, the larger the piece of meat being cured, the longer it will take to complete the curing process. The amount of salt used can be too little or too much, which means it's important to follow the instructions closely. If too little salt is applied, bacteria can grow and spoilage can follow. If too much salt is used, the meat can become hard, dry, and taste overly salty.

The term *cure* can be used to describe either the actual process of turning raw meat into a preserved, safe, and edible meat or the commercial product pack containing the essential ingredients, such as salt and sodium nitrite, that cause meat's transformation from raw to preserved.

*Cure* is sometimes interpreted to mean both curing and the subsequent smoking of meat. However, curing does not actually imply smoking, although the two processes work well together. Strictly speaking, curing applies only to dry salt curing, brine submersion, or pickling with a vinegar base. In a wider sense, curing applies to any saline or alkaline preservation solution, perhaps with some modifications.

**NOTE:**
The days when meat for eating was cured by drying it in the sunshine and open air are a thing of the past. While a traditional method, it is no longer recommended as a viable option for meat preservation. Outdoor drying creates unstable temperature and moisture variations (and can lead to insect infestations), all of which can lead to the growth of spores causing botulism.

If you plan to make smoked, cooked or dry sausages, or jerky, you will need a cure product. Today's alternatives are much better and more stable than the saltpeter (a strong form of nitrate—sodium or potassium) typically used by earlier generations of meat preservers. Why use a cure?

It's because of the bacteria that can be produced by decomposing muscle. *Clostridium botulinum* is the bacteria that causes botulism, a potent and deadly form of food poisoning. The spores thrive in meat environments with temperatures between 40°F and 140°F (4°C to 60°C) in moist, low-oxygen conditions. This is exactly the environment we provide in sausage smokers, dry sausages, and fresh pork or other meats held at room temperature.

At this time, there is no known substitute for nitrite in curing meat and sausages. The benefits of using it far outweigh any health risks that may be associated with it. You can go without using nitrites if cooking fresh ground beef, pork, lamb, venison, or other meats, but sooner or later, you will need to consider how best to preserve the rest of the fresh meat you can't use right away.

The botulism-causing bacteria are present in many soil conditions, vegetables, and other foods we consume. So, how real is the danger of it? If you consume the bacteria, it is a very real danger, and it comes with a high risk of dying or severe nerve damage. Yet it is not likely you will ever experience its effects because commercially available food products are strictly regulated and monitored. Most botulism cases occurring in the United States are the result of improper home canning. Botulinum spores are hard to kill but aren't harmful except, potentially, to infants. The spores in the soil and vegetables typically are not found in sufficient amounts to be deadly to humans. However, when these bacteria are allowed to grow in an oxygen-free (anaerobic), non-acidic environment between 40°F and 140°F (4°C to 60°C), they will multiply rapidly and start producing the deadly toxin.

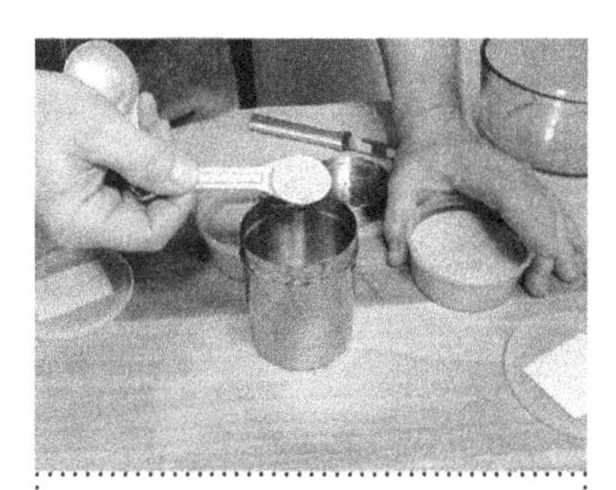

Curing salts are required if making sausages or jerky. They help prevent the growth of unwanted bacteria in the 40°F to 140°F (4°C to 60°C) window, where bacteria can grow rapidly. Commercial cures are readily available, and you should follow the exact instructions for their use.

## CURING SALTS

To inhibit botulinum spores and their growth, a curing salt of some form *must be used* in any dry-cured sausages. There are no exceptions.

Many family recipes handed down through the generations used saltpeter, or potassium nitrate, for curing. Most meat supply companies no longer sell saltpeter, but you may find other commercial products that will achieve similar results. There are commercial fast-cures available that contain 0.5 percent sodium nitrate and 0.5 percent sodium nitrite. If used, they are typically found in recipes at a ratio of 1 teaspoon (0.17 ounces or 7 grams) per 1 pound (455 g) of meat.

Sodium nitrite, often referred to as pink salt because of its color, prevents these bacteria from growing. Sodium nitrate will act as a sort of time-release capsule form of sodium nitrite and must be used in all dry-cured sausages cured for long periods, such as salami, which may be cold-smoked and then dried for weeks.

Be aware that these cures themselves can be dangerous if ingested, such as if you accidentally lick a finger covered with these salts. There is a reason for curing salts to appear in recipes, but you should always use them in the proportions stated in the recipe. Don't alter recipes with curing salts, don't try any of the salts to see what they taste like, and keep them out of the reach of children!

While you need to take precautions when using curing salts, they are beneficial and have three main functions: killing a range of bacteria, especially those responsible for botulism; preserving the pink color we associate with meats; and adding a tangy flavor to the meat.

Nitrates do nothing beneficial to food until they convert to nitrite. Potassium nitrate (saltpeter) was used in curing until the 1970s, after which it was largely discontinued because it was too inconsistent to be safe. Sodium nitrate is now manufactured and sold under the commercial brand names of InstaCure #2 and DQ Curing Salt #2.

For purposes of discussion rather than recommendation, three commercially available brand-name cures are frequently used: Prague Powder #1 and #2, Morton Tender Quick, and InstaCure #1 and #2. Another cure is tinted curing mix (TCM), which is also referred to as Prague powder or pink salt.

Regardless of the name, the composition is generally the same: 93.75 percent salt and 6.25 percent nitrite. However, always use them based on the directions given by the supplier, as one brand may have a higher concentration of sodium nitrite than another.

As an example, Prague Powder #1 and #2 are used for different products: #1 is for most cured meats and sausages, except for cured meats such as salami, and #2 is used for dried meat and sausages. Both are used in very small quantities, but, again, you must follow the supplier's directions exactly. Morton Tender Quick contains both sodium nitrite and sodium nitrate. It has a lower nitrite-to-nitrate concentration (0.5 percent of each) and much more salt than the other cures.

This makes the Morton product good as a rub or in a brine, but it has a more limited use in sausage making because, with the extra salt, it can get very salty before the correct amount of cure is reached. InstaCure #1 and #2 are similar to the Prague Powder and can work very well for any sausage making or meat curing. You should do your own research before sausage making to make the decisions that will affect your resulting products.

The bottom line is that *you need to use a cure if you want safe smoked, cooked, or dried sausage or other meats.* Sodium nitrate, and the sodium nitrite it produces, is a safe product for curing meats and sausages.

## DRY RUB CURES

A dry rub cure consists of a mix of dry ingredients including common salt and other preservatives, as found in commercial cures, that is sprinkled on and then rubbed into the meat. The ingredients are then allowed to work into the meat before it is dried. Salt is often referred to in recipes as "sodium chloride." "Commercial cure" often refers to salts that have nitrite in them and sometimes nitrates. Dry cures typically contain herbs and spices, and some contain sugar or other sweeteners.

If you are using a dry rub, it is best to allow twenty-four hours for it to work into the meat. This will require that the meat be put in a pan, covered, and left in your refrigerator until ready for use.

If you sprinkle a cure on the meat, use a sugar shaker that has large holes.

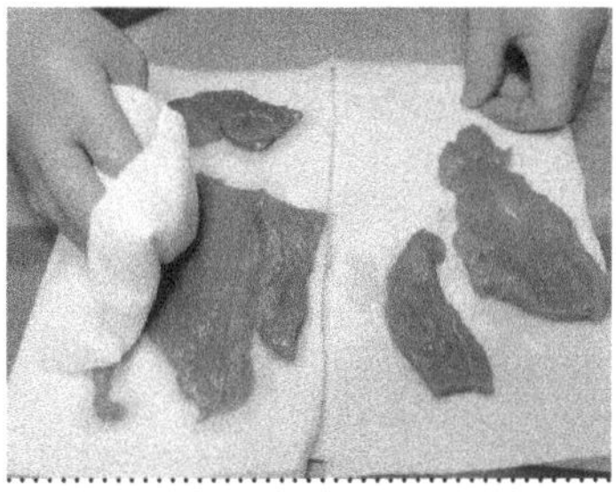

Once you've marinated the strips, remove them from the marinade, place them on paper towels, and pat them dry. Then, place them on a dehydrator tray or oven rack for drying.

## NOTE:

The USDA has designated the "safe" level of nitrates or nitrites to be 200 ppm. This figure represents the grams of nitrate and nitrite times 1 million divided by the grams of cured meat that it treats. For example, 200 ppm of nitrate for 50 grams (2 ounces) of cured meat is equal to 0.01 grams of nitrate ($0.01 \times 1$ million $\div 50 = 200$ ppm). This level preserves the antimicrobial power of these compounds while preventing the development of carcinogenic concentrations. Remember, it is difficult to remove nitrates or nitrites from the meat-curing process without increasing the risk of harmful bacteria, particularly botulinum.

# SODIUM NITRITE AND NITRATE CONCERNS

Nitrates and nitrites are chemical compounds that contain nitrogen and oxygen and are commonly used in curing meat. They have been used in place of salt for many years and are typically added in the first step of any commercially cured product to inhibit bacterial growth. Their action removes the moisture that bacteria could live on and kill the bacteria through dehydration. Sodium nitrate is particularly effective in food preservation and is widely used because it allows the food to last longer.

You should be aware of the health concerns that sodium nitrate and sodium nitrite have posed for many people. At too high of a level, they can be detrimental to health.

Concerns about consuming nitrates or nitrites center on the quantity eaten rather than their inclusion as preservatives. A University of Minnesota report concluded that nitrite as it is used in meat such as sausages is considered safe because the known benefits outweigh the potential risks. Fresh sausages are the only type you can make if you are concerned about nitrites and nitrates in your diet.

A small barrel smoker has two compartments. One holds the wood (left), and the other is a chamber for placing the meat to be smoked. The small unit shown can be easily used and moved around your yard.

Sensitivities to nitrates and nitrites can cause migraine headaches or allergic reactions for some people. At high enough levels, these compounds can be carcinogenic, and a large enough dose can be lethal. The University of Minnesota published findings in 1992 that a fatal dose of potassium nitrate, or saltpeter, is in the range of 30 to 35 grams (slightly more than 1 ounce) for adults, consumed in a single dose. The fatal dose of sodium nitrite is in the range of 22 to 23 milligrams per kilogram of body weight.

To reach a lethal toxicity level, an adult weighing 150 pounds (68 kg) would have to consume about 20 pounds (9 kg) of brine-cured meat containing 200 parts per million (ppm) nitrite in one meal. (Even if a person could eat that amount of meat in one sitting, it is likely that the salt level, not the nitrite, would create the toxicity.) Keep in mind that the normal American diet contains more nitrates from leafy vegetables such as celery, spinach, radishes, cabbage, beets, and lettuce than from cured meats because these plants readily absorb nitrogen fertilizers used in food production. In short, unless your daily diet consistently includes high volumes of cured meats, this is not likely a topic you should lose sleep over.

Not specifically referred to as a cure, Fermento is a commercially available dairy-based product made from cultured whey protein and skim milk that helps in the fermentation process. It is used to produce a tangy taste in semidry sausages, such as some summer sausages and Thuringer sausage. The "tang" found in fermented, dry-cured meat is due to a decrease in pH as the lactic acid builds up. This product mimics that taste, as does citric acid. However, citric acid is not lactic acid and will not yield the same flavor. The key to a tangy-flavored sausage is proper fermentation produced by specific bacteria that are added to the meat as a starter culture. You will have more control over this flavor by adding the recommended amounts.

The recommended level of use based on current formulations is 3 percent, or about 1 ounce (28 g) per pound (455 g) of meat. It is possible to double this percentage to produce a more tangy taste, but if you exceed 6 percent, the sausage likely will become mushy. Fermento does not need to be refrigerated and quickens the fermentation process. Instead of the several days that is often required for starter cultures to start fermentation, once this product is added, the fermentation will take only hours before you can begin smoking the meat.

# THE BASICS OF SMOKING

The purpose of smoking meat when preserving is to lower the moisture, which reduces the opportunity for bacterial growth. Smoking is not the same as grilling; it gives meat a unique smoke flavor and often provides a more attractive external color than only curing. Country-cured hams and similar cured and smoked meats that do not require refrigerated storage owe their stability to a combination of low moisture, high concentrations of curing agents, and heavily smoked surfaces. Three factors affect the amount of time a meat product needs to be smoked: the type of meat product, the density of smoke generated within the smoking unit, and the ability of the meat surface to absorb the smoke properties.

Hot smoking and cold smoking are the two methods for smoking meat, and they are identified by the temperatures used with them. These temperatures will be applied to different meats in several ways, though in all cases, the meat still must reach its critical internal temperature to be safe to eat. These internal temperatures will vary slightly between animal and fowl species. Pork, for example, needs to reach an internal temperature of 160°F (71°C) to be considered safe, while poultry requires 165°F (74°C).

When smoking meat, you need to keep in mind at all times what is referred to as the "temperature danger zone." This zone, between 40°C to 140°F (4°C to 60°C), is the range that provides an environment for rapid microbial growth. You will need to ensure that proper steps are taken to eliminate all risks involved.

Hot smoking uses heat and smoke together to make a product that is tasty and safe to eat. You are cooking and flavoring the meat at the same time. Hot smoking is the most common method used for smoking meats and is the recommended process for those starting out. It creates temperatures above 145°F (63°C), past the danger zone. Typically, hot-smoking temperatures range between 165°F to 185°F (74°C to 85°C) and may increase to 200°F to 300°F (95°C to 150°C) if needed (for specifics, see the chart on page 152).

Hot-smoked meats are usually smoked in the same chamber as the burning wood. During this process, he meat is cooked with heat created by the gases of the fire, and the smoke provides the flavor. A slow smoking process better allows the smoke to penetrate the meat and helps tenderize the meat as well.

No matter your level of familiarity with your smoker, it will require attention and patience to control several variables when hot smoking. For example, to provide a good ventilation of the smoke while still maintaining the proper heat, you need to make sure the smoker has proper airflow. Also, while it might seem like a contradiction, it may be necessary to maintain a degree of humidity inside the smoker through the dehydrating process to prevent the meat from getting too dry or too smoky.

Cold smoking smokes meat without exposing it to high heat. This process, when done correctly, can take many hours or even days. Because of this lengthy process, the end result typically is a drier meat and likely saltier due to the intensity of the brine, cure, or salts used. On the plus side, for those who are skilled at using this process, it can yield complex flavors.

As indicated by its name, cold smoking requires a lower temperature, usually below 90°F (32°C). With this temperature, the meat will take on a smoke flavor while remaining fairly moist.

You should be aware, however, that cold smoking *does not* cook the meat. Whatever meat you are cold smoking *must be fully cooked* before you begin this process. The reason for this is that you are holding the meat in the temperature danger zone for an extended period of time. Failure to properly process meats with cold smoking can cause severe health problems or even death.

The main pathogens that can be created are *Clostridium botulinum*, which creates toxins that can kill people or severely damage or disable the immune and nervous systems. *Listeria monocytogenes* is another pathogen that is lethal and can grow in underprocessed meats and fish.

The cold-smoking process is not recommended for newcomers to smoking meat and is not covered in this book. Use the hot-smoking method instead.

## AFTER THE SMOKE CYCLE

After the smoke cycle is complete, you can gradually increase the temperature inside the chamber to cook the meat. Avoid a rapid increase in temperature, as this will dry and overcook the surface before the desired internal temperature is reached. Increasing the temperature in increments will conduct the heat through the meat to minimize the difference between the surface and internal temperatures. A long, slow cook of the meat at this point will tenderize it to maximum effect.

If possible, try to cook meat pieces that are a similar size, as this will allow you to cook them at a specific temperature for an equal time and have a uniform result. Unevenly matched pieces either may be overdone and too dry or undercooked, depending on temperature and time. You may be able to circumvent these problems by using individual temperature probes to determine when target internal temperatures have been reached and the cooking cycle should be stopped. If this proves impractical for you, several cooking sessions may be needed. As a rule, high smoke-unit temperatures (110°F [43°C] and above) with a light smoke will speed up the drying process while lower temperatures (80°F to 110°F [27°C to 43°C]) with a dense smoke will intensify the smoky flavor in meat.

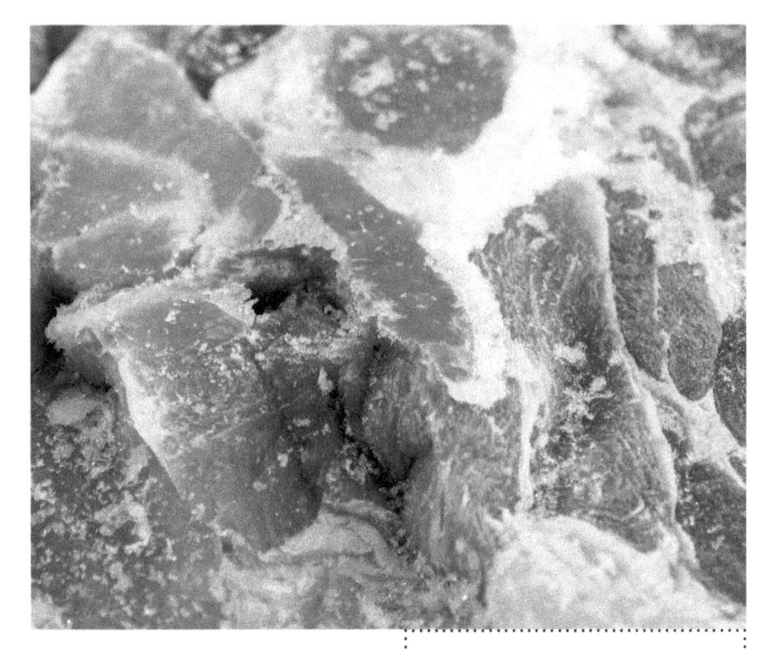

The type of cured and smoked meat products you want to produce will determine the level of the smoking-unit temperature. If you are storing meat at air temperature, it should be smoked at a temperature of 135°F (57°C) until the inside of the meat reaches 110°F (43°C). You can then lower the unit's temperature to 110°F (43°C) and maintain temperature until the desired color is reached.

Remove the meat after it has been smoked and wash it to remove salt and fat streaks from the surface. Cooked meat products should be cooled quickly to 40°F (4°C) or below. At this point, it is important to maintain sanitary conditions and avoid contact with uncooked meat or surfaces that have come in contact with uncooked meat. This will minimize recontamination of the cooked products with organisms that may cause spoilage.

Frozen meat needs to be thawed completely before smoking. Always thaw frozen meat in your refrigerator and never at room temperature. Also, never refreeze meat that has been thawed, unless it has been cooked.

# SMOKING FROZEN MEAT

You can smoke meat that has been previously frozen, but you must first thaw it out completely before smoking it. Because smoking uses low temperatures to cook food, the meat will take too long to thaw in the smoker, allowing it to linger in the danger zone between 40°F and 140°F (4°C to 60°C), where harmful bacteria can multiply.

Never thaw meat at room temperature. Keep it cold in your refrigerator while it is thawing. This is essential to keep harmful bacteria from growing while it is thawing. You can microwave the meat to thaw it more rapidly, but you then must smoke it immediately because parts of the meat may begin to cook during the microwave thawing process.

You can also thaw the meat by wrapping it in an airtight package and submerging it in cold water. Change the water every thirty minutes to maintain a temperature below 40°F (4°C). When thawed, it must be cooked immediately.

# THERMOMETERS

You will need two types of thermometers to make sure the meat is smoked safely. You'll want one thermometer to monitor the air temperature in the smoker to be sure the heat stays between 225°F and 300°F (107°C to 150°C) throughout the cooking process. Many new smokers have built-in thermometers to help with this.

The second thermometer is to determine the internal temperature of the meat. Oven-safe thermometers can be inserted in the meat and remain there during smoking. Once the meat is removed from the smoker, you can also use an instant-read thermometer to double-check the temperature.

## CALIBRATING YOUR THERMOMETERS

Before you move on, calibrate your thermometers. This may seem tedious, but it is important. Without an accurate thermometer, any readings you take will lead to wrong assessments of temperature. You should calibrate thermometers before every use and whenever they are dropped. They are sensitive and can lose accuracy from extensive use or when going from one temperature extreme to another. Remember that you are using them with meat, and incorrect cooking temperatures can lead to undercooked meat, which can pose a health risk.

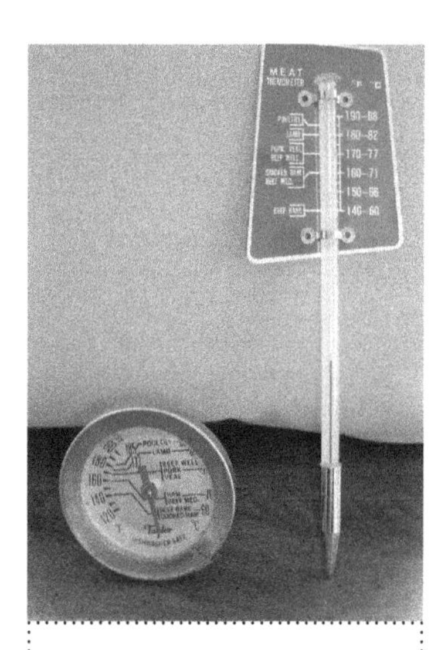

Accurate temperature measurements are critical for cooking and/or smoking any kind of meat. Many varieties of thermometers are available, including instant-read and digital, that can be inserted into the meat. Some types available can be read outside the smoker. Those with small probes are easier to use for monitoring meat used for jerky.

Thermometers can be calibrated by the use of ice water or boiling water. In the ice-water method, use these steps:

1. Fill a 2-cup (475 ml) measuring cup with crushed ice and water and stir well.
2. Let sit four to five minutes to reach 32°F (0°C).
3. Completely submerge the sensing area of the thermometer's stem or probe in the ice water, but keep it from touching the sides or bottom of the container.
4. Hold for thirty seconds, or until the displayed temperature stabilizes.
5. If the thermometer is not within plus or minus 2 degrees of 32°F (0°C), adjust the thermometer accordingly. The ice-water method permits calibration within 0.1 degree. (Some digital stemmed thermometers have a reset button, which makes calibration especially easy.)
6. Repeat the process with each thermometer.

For the boiling-point method, follow these steps:

1. Fill a saucepan with purified or distilled water and bring to a rolling boil.
2. Completely submerge the thermometer's stem or probe in the boiling water without touching the sides or bottom of the pan. Wear an oven mitt if your hand will be close to the surface of the boiling water.
3. Hold for thirty seconds, or until the displayed temperature stabilizes.
4. If the thermometer is not within plus or minus 2 degrees of 212°F (100°C), adjust until it is. The boiling point method permits calibration to within 1 degree. Repeat the process with each thermometer.

Note that the boiling point will be affected by altitude. The boiling point of water is about 1 degree Fahrenheit lower for every 550 feet (168 m) above sea level. If you are in high-altitude areas, adjust the temperature by calibration. For example, if you are at 550 feet (168 m) above sea level, the boiling point of water would be 211°F (99°C).

Any food thermometer that cannot be calibrated can still be used by checking it for accuracy using either method described. You can take into consideration any inaccuracies and make adjustments by adding or subtracting the differences, or consider replacing the thermometer.

Jerky can be made from many different kinds of meat, whether domestically raised or taken from the wild. Using thin strips provides the best result after the curing, cooking, smoking, and drying processes have been completed.

# CURED AND SMOKED PRODUCTS

A vast variety of flavors and textures can be produced from smoking and curing different meat cuts. Pork, beef, poultry, and wild game can be smoked and cured. The list includes but is not limited to the products below:

*Bacon*: Pork bellies are usually trimmed into a rectangular shape and smoked and cured before being sliced into strips. Although it has been smoked and heated, bacon still must be cooked before it is consumed. Cooking methods include frying, grilling, broiling, or microwaving.

*Dried beef*: The beef round, sirloin tip, and larger muscles of the chuck may be used to make dried beef. Dried beef has a lower moisture content than many other beef products. It is smoked to varying degrees but is fully cooked and ready to eat after curing.

*Jerky*: These are dried meat strips that may be produced using a combination of curing, smoking, and drying. After drying, jerky does not need refrigeration if it is packaged to prevent it from absorbing moisture and contamination from its surroundings during storage.

*Ham*: Hams are a popular smoked meat product that may be boneless or boned. Because they are larger than other meat cuts, hams will take longer to cook, cure, and smoke.

# HOT SMOKING: COOKING TIMES AND TEMPERATURES

| TYPE OF MEAT | WEIGHT (LB/KG) | SMOKER TEMP. (°F/°C) | COOKING TIME (HOURS)* | FINISHED INTERNAL TEMP. (°F/°C)** |
|---|---|---|---|---|
| Beef Brisket (Pulled) | 10 to 12 lb (4.5 to 5.5 kg) | 225°F to 240°F (107°C to 116°C) | 1.5 hr per lb (0.5 kg) 12 to 20 hours | 190°F (88°C) |
| Beef Brisket (Sliced) | 10 to 12 lb (4.5 to 5.5 kg) | 225°F to 240°F (107°C to 116°C) | 1.5 hr per lb (0.5 kg) 12 to 20 hours | 200°F (95°C) |
| Beef Chuck Roast | 3 to 5 lb (1.5 to 2 kg) | 225°F to 240°F (107°C to 116°C) | 8 to 12 hours | 200°F to 210°F (95°C to 99°C) |
| Pork Butt (sliced) | 6 to 8 lb (3 to 4 kg) | 225°F to 230°F (107°C to 110°C) | 1.5 hr per lb (0.5 kg) 14 to 16 hours | 200°F to 205°F (95°C to 96°C) |
| Pork Ham | 12 to 16 lb (5.5 to 7 kg) | 170°F to 180°F (77°C to 82°C) | 6 to 8 hours | 145°F (63°C) |
| Pork Shoulder | 6 to 8 lb (3 to 4 kg) | 225°F to 230°F (107°C to 110°C) | 12 to 14 hours | 200°F to 205°F (95°C to 96°C) |
| Bacon | 2 lb (1 kg) | 175°F to 180°F (79°C to 82°C) | 2 to 3 hours | 150°F (66°C) |
| Spare Ribs | 2 to 3 lb (1 to 1.5 kg) | 225°F to 240°F (107°C to 116°C) | 6 hours | 180°F (82°C) |
| Bratwurst | 2 to 3 lb (1 to 1.5 kg) | 225°F to 240°F (107°C to 116°C) | 2 hours | 165°F (74°C) |
| Chicken (Whole) | 4 to 5 lb (~2 kg) | 250°F to 275°F (120°C to 140°C) | 3 to 4 hours | 165°F (74°C) |
| Turkey (Whole) | 12 to 14 lb (5.5 to 6.5 kg) | 240°F (116°C) | 6 hours | 165°F (74°C) |
| Chicken/ Turkey Breasts | 1 to 2 lb (0.5 to 1 kg) | 225°F (107°C) | 3 hours | 165°F (74°C) |
| Quail/ Pheasant | 1 to 2 lb (0.5 to 1 kg) | 225°F (107°C) | 2 hours | 165°F (74°C) |
| Salmon Fillets | 2 to 3 lb (1 to 1.5 kg) | 160°F (71°C) | 4 hours | 130°F (54°C) |
| Whole Trout | 1 lb (0.5 kg) | 225°F (107°C) | 1 hour | 145°F (63°C) |
| Meatloaf | 2 to 3 lb (1 to 1.5 kg) | 225°F to 240°F (107°C to 116°C) | 3 to 4 hours | 160°F (71°C) |
| Venison Steaks | 1 to 2 lb (0.5 to 1 kg) | 210°F to 220°F (99°C to 104°C) | 3 to 4 hours | 160°F (71°C) |

*Cooking times are approximate depending on thickness of cut. Use an estimate for wild game and other cuts from other species that use 1.5 hours per pound (0.5 kg) of meat.

**Always use an accurate thermometer to check the internal temperature. Time is an estimate of doneness and not an indicator.

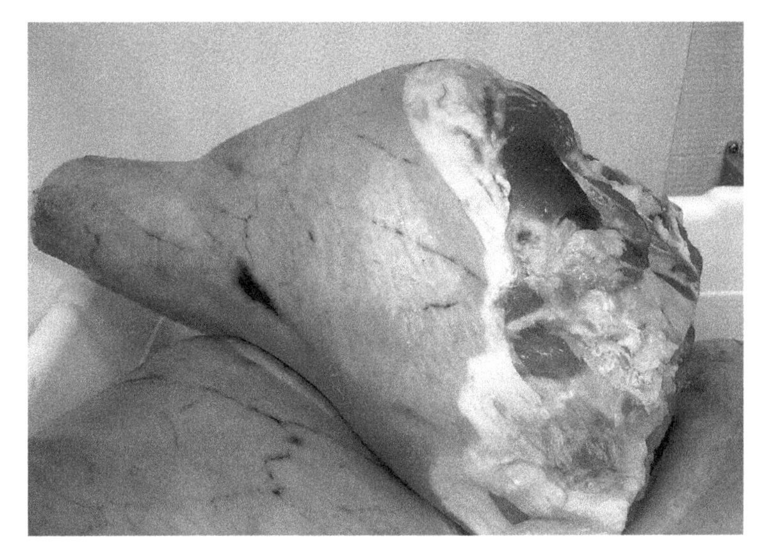

Smoking whole hams is possible, but it requires more time and attention than if the meat is cut into smaller pieces.

*Poultry*: You can cure and smoke whole birds by controlling the smoke intensity and duration, both of which will affect the final smoke flavor. Hollow birds will cook more rapidly than solid pork hams or other dense cuts. Ducks and geese have larger amounts of carcass fat than other birds, so you should trim as much fat as possible to reduce the amount of grease that will drip away. Fat from these birds has a low melting point, and it may streak the smoked surface during the cooking phase.

*Wild game*: Most wild game animal and fowl meats can be cured and smoked. Apply salt, spices, seasonings, and other ingredients to suit your tastes.

## SMOKING PORK

Any cut from the carcass can be smoked, and this section will guide you through the smoking processes. If you choose to smoke pork cuts, you should be aware of the possible presence of trichinae, a small slender worm that may be present in the muscle. It is a parasite when in a larval stage in the voluntary muscles of humans and hogs. An infection occurs after the pig ingests the parasites. They burrow into the muscles and can cause muscle pain, fever, and other physical effects.

Not all pork will be infected by these parasites, whether you purchase your pork at a market or raise the animal yourself. However, because these parasites are too small to be seen without using a microscope, you need to be cautious and take several steps to eliminate any potential problems.

First, cook all fresh pork to a minimum internal temperature of 145°F (63°C) and allow a three-minute rest after cooking. During this rest time, the temperature remains constant or continues to rise, ensuring the destruction of any harmful germs. The internal tissue does not stop cooking at the exact moment you remove the meat from the heat source. The internal heat will continue to penetrate into the tissue until the heat source is removed. Once it is removed, there is no longer a heat penetration into the tissue, and the internal temperature will start to drop after about three minutes.

Previous cooking guidelines for pork called for a minimum internal temperature that was higher than 145°F (63°C). The new recommendation reflects advances in food safety and the nutritional content of pork. Research shows that most common cuts of pork are 16 percent leaner than twenty years ago, and the saturated fat has dropped 27 percent. This drop in pork muscle fat content allows for a lower cooking temperature to reach the safe degree of doneness.

Almost every part of the pig can be smoked. Some cuts are larger than others and will take more time to smoke, but all can be used. Bacon, hams, and shoulders are the cuts most often smoked. Jowls, ribs, and loins can be smoked too. Even pork sausages can benefit from some time in the smoker.

Bacon, one of the most popular smoked products, comes from the belly section of the carcass. If meat from other portions of the carcass is used, it may carry the name of where it came from, such as pork shoulder bacon. Pork bacon must be cooked before eating. Most bacon made from your pig carcass will be streaky bacon—the long narrow slices cut crosswise from the belly that contain veins of pink muscle layered within the white fat. These muscles and surrounding fat hold a pig's internal organs intact and protect it from outside injury. You will have trimmed the belly into rectangular or square shapes before slicing into strips. Although you may smoke and heat the bacon, it must be cooked before it is eaten.

Hams are popular for smoking, either boneless or with the bone. A ham that has the center bone removed will take less time to heat and smoke than one in which the leg bone is still present. It will take the heat longer to penetrate completely through to the bone and reach the critical 145°F (63°C) next to the bone. When smoking a ham, it is best to remove the skin and fat, as the smoke will not penetrate into the meat if it is still on. And when that outer layer of skin and fat is removed, all smoke flavor will be removed too.

Pork shoulders can be smoked, and as the name implies, they are located in the front part of the carcass. These are large cuts and, like hams, will take longer to smoke and heat unless they are cut into smaller pieces. Ribs, jowls, and loins are smaller cuts that can be smoked and will take less time to finish. The term *Canadian bacon* refers to round slices of pink meat from the loin.

Try to cook pieces that are similar in size, as this will allow you to cook them at a specific temperature for an equal time and have a uniform result. Unevenly matched pieces may become overdone and too dry or undercooked and unsafe.

## SMOKING BEEF

Beef shoulders, briskets, roasts, loins, ribs, and rounds are some of the more popular parts of a cow to smoke. However, pretty much any part of the animal can be smoked if desired.

The largest parts include the shoulders and rounds, or the hindquarters, and the brisket, which is that portion located in the lower front of the chest and between the front legs. These are the most-used muscles of the cow's body because they propel the animal's movements. The high use of these muscles creates a leaner and more fibrous texture when cooked. Their location also means there is a different fiber configuration within the muscle, and they contain ligaments, which attach the muscles to the bones.

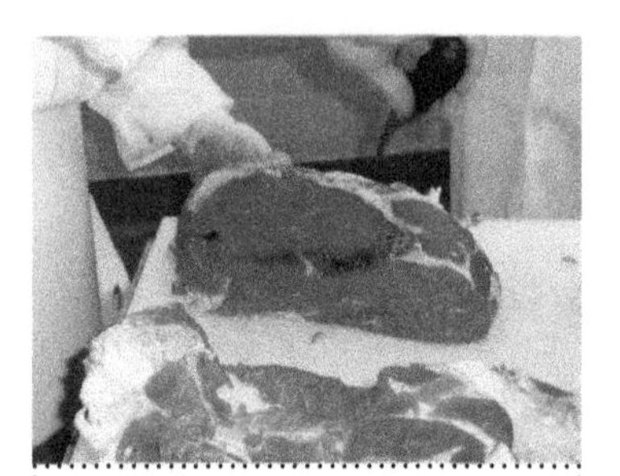

A beef round can be smoked and dried with great results. When the process is correctly completed, the round can be cut into slices for sandwiches, hors d'oeuvres, or party plates.

As with other cooking processes, the larger the cut of meat, the longer it will take to smoke completely. For example, a beef brisket should generally be smoked for about 1.5 hours per pound (455 g) of meat. Depending on the size of the cut, this may require anywhere from twelve to twenty hours of smoking time. For these larger cuts, you may want to divide the meat into several smaller, equal-size pieces. (If you mix larger and smaller pieces in the smoker, you will either overcook the small pieces or undercook the large pieces.) Although time is a good guide, the internal temperature of the meat is the most important indicator of doneness. Having an accurate meat thermometer will help you monitor internal temperatures.

Smoking venison is an option for the rounds, loins, or shoulders of the animal. Be sure to reach a minimum internal temperature of at least 160°F (71°C) for any large wild game you want to smoke.

For most purposes, an internal temperature of 200°F (95°C) will be your goal. This will kill any organisms that may be present in the muscle and allow for safe eating and storage.

Just as you do with other cuts and other species, such as pork, be sure to remove as much of the outside fat as possible before you begin the smoking process. The rounds, shoulders, and roasts should be handled in the same manner.

Beef ribs are typically larger than pork ribs. Low and slow is the key to smoking beef ribs in order to make them tender enough to pull easily off the bone. Again, depending on the size of the ribs, the average cooking time will generally be about four to six hours. There will be other factors affecting this time, including the type of smoker you use. Monitoring the progress of the smoking process and taking good notes will help you achieve a satisfying end result in future smoking sessions.

## SMOKING WILD GAME

Smoking wild game is similar to smoking meat from a domestically raised animal. You must reach the target temperatures with cooking and/or smoking to kill any bacteria or parasites that may inhabit the meat. However, there are a few important differences that you should keep in mind.

First, the animal is, in fact, wild. Any animal or fowl you take from the outdoors will not have had the advantages of being fed a steady or controlled diet. It will also not have received any attention relating to health protocols, including vaccinations, parasite control, and government inspection. This means you will be determining whether the animal or bird in question is a healthy one to use for your table.

Generally, animals caught in the wild are much leaner than those bought in stores. The lean diet from foraging, plus the exercise these creatures get while on the move looking for food, creates a body that has less fat. Because there is less fat, wild game meat will more readily absorb more of the smoke flavor than a domestic animal's meat. This can result in an "overdone" smoke character that leaves a bitter flavor. You may want to adjust wood choice and the amount of airflow to counteract this.

Because wild game meat is leaner, it also may be described as "tougher" and not as pliable. Smoking will loosen the muscle fibers and help tenderize the meat somewhat, but you will need to monitor your smoking process for the best results. Wild game will have a different flavor than commercial meat no matter how you cook it. It typically has a stronger taste and while some like this dynamic, others don't. Be aware of this difference and know that a good or bad flavor might not necessarily come from the smoking process.

Electric smokers can provide a uniform and consistent temperature for a specified period of time. They are popular because they are easy to operate, take up little space, and are reliable when properly used.

In order to properly smoke your wild game or fowl, you will need to weigh the pieces you want to smoke. First, trim away as much of the fat as you can and then weigh the meat. This will help you determine the amount of seasoning and cure you will need. Quality recipes usually have their seasonings stated by the pound (grams or kilograms).

The key to smoking wild game is to do it low and slow. While a higher temperature results in a shorter smoke time, it affects the end quality and shelf life. The longer a piece spends in the smoker, the more moisture it loses and the saltier it becomes. Both of these factors help preserve it better. If you plan to eat the meat right after smoking, a higher temperature can be used. Whatever you do, always reach the temperatures required to kill all bacteria and parasites.

## SMOKERS

Smokers are available in a tremendous variety of prices, sizes, and functions. They range from home grills with covers to substantial upright units that are a large fixture in a backyard. They can be electric or fully powered by wood. If you want only to add a smoky flavor to your meat, you can use just about any model—even a home grill with a smoke box or smoking attachment will work just fine. However, if you are smoking to preserve meat in conjunction with curing, you will want more precise temperature control.

Smokers specifically designed for meat do two things at once: they provide a proper temperature to kill harmful pathogens and a pleasing smoky flavor. The heat provided can originate from several sources including electricity, wood chips or pellets, gas, or charcoal briquettes. If you're using wood, different varieties can be used to infuse a unique smoke flavor or finished color (see page 160).

The amount of meat and the size of the cuts will largely determine the type of smoking unit you'll find most useful, although your budget and the amount of space you can dedicate to a smoker will also need to be taken into consideration.

The most commonly available home smokers are vertical electric water smokers, insulated variable-temperature smokers, electric smokers, and stovetop smokers. Understanding their advantages and limitations may help you decide which is best for your situation. While these are brief descriptions, containing the unendorsed mention of only a

A closed grill can be used to smoke small amounts of meats and sausages. Be sure to use an oven-safe thermometer to monitor the internal temperature of the meat as it cooks.

Large barrel smokers can hold more meat and more wood for the smoking process. Like any other smoking unit, the heat level needs to be monitored the entire time to produce the best results.

**NOTE:**
One key concern when evaluating the design of a smoker is the potential loss of heat caused by opening a door to add water to a pan or to replenish wood chips, pellets, or logs. Models are available that have external wood chip or pellet loaders so the unit doesn't need to be opened. Some smokers have a tray that can be pulled out, have fresh chips or pellets added to it, and be pushed back into the unit without opening the door.

few models to help illustrate the smoker types, you are encouraged to research all the models that are commercially available to determine which one may work best for you. Prices range from modest to expensive, and this may give some indication of their ability for extensive use. Some are stationary, while others may be set on rollers for easier movement. They may be rectangular or round, and the number of trays will increase as the unit expands in size. Many different types of digitally controlled smoker models are available and popular because they can be used to easily monitor the temperature and time.

## SMOKER TYPES

*Vertical water smokers* are popular because they are generally the least-expensive smokers on the market. However, the less-costly models may not reach the high temperatures you need. The more expensive units, such as the Weber Smokey Mountain, have better temperature control. These units have either a gas or electric heat source and typically have three components: a bottom heat source, a water pan that stores heat and regulates the internal temperature, and a smoking chamber. The biggest disadvantage is the loss of heat when the lid is opened. You can mitigate this by having a thermometer that can signal the temperature to an outside receiver.

*Insulated variable-temperature smokers* have good temperature control. This variety is becoming more popular with those who want to do home smoking. They are typically more expensive than other models but are easy to use and generally similar to a vertical electric water smoker.

*Electric smokers* are another popular type because they are easy to use and don't take up a lot of space. The more expensive models typically have a rheostat that turns down the electricity flow to the coil, much like those found on an electric stove or hot plate, and they may have multiple settings ranging from low to high. Some of the more expensive electric smokers have thermostats with a temperature probe inside the cooking chamber. The thermostat monitors the temperature and will raise it if it's too low or lower it if it's too high. This makes a unit with a thermostat better than one with a rheostat, but it also makes the unit more expensive. One drawback to this type of unit is that it doesn't work well outdoors in cold weather.

A simple smoking chamber can be made by digging a firebox pit, using heat-resistant piping, and setting a metal chamber above the entrance of the smoke pipe. One reason to bury the pipe is to keep it from creating a fire hazard that might ignite leaves, dry grass, or other refuse surrounding it.

The principle for a homemade smoker is to allow the smoke from the wood to rise to a separate chamber where the meat is suspended. You can build your own unit, but be aware of any local ordinances before building a permanent structure.

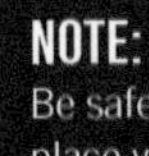

## NOTE:

Be safe! Depending on where you place your smoking unit, make sure there is adequate ventilation so that any escaping heat and smoke does not create air-quality problems, such as carbon monoxide in your home, shed, or apartment. Carbon monoxide is an invisible, odorless gas that can be produced by malfunctioning appliances, such as gas- or wood-burning stoves, fireplaces, and smokers. Carbon monoxide alarms are available, and you should have one installed inside your home if you use a smoker indoors.

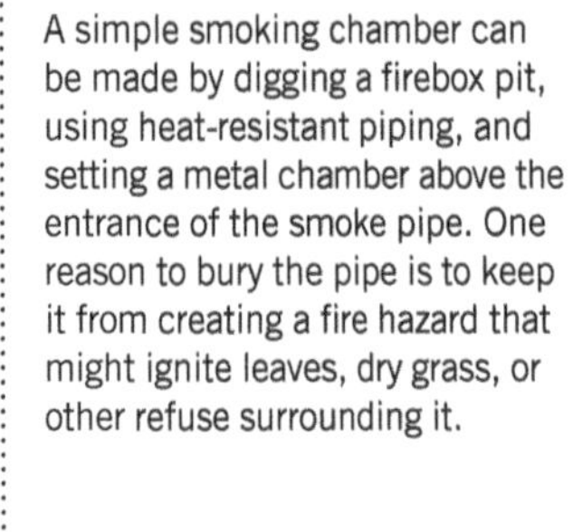


*Stovetop smokers*, such as the Cameron stovetop models, have become available in recent years. Stovetop smokers work well in apartments or places where other smoking units can't be used. They are inexpensive, easy to use and clean, and will work with a variety of meat including beef, fish, poultry, wild game, and waterfowl. However, since most of these models range between 7 and 11 inches (18 to 28 cm) wide and 11 and 15 inches (28 to 38 cm) long, they are limited in the amount of meat they can hold at one time. As for design, the stainless-steel unit is an enclosed system that uses your stovetop for heat to activate the flavored wood chips sprinkled across the inside bottom of the pan. A grill rack is set above the base, on which the meat is placed. The cover tightly seals in the heat and smoke.

*A covered grill and charcoal-fired smokers* can be used to smoke meat, although it is more difficult to maintain an even temperature and smoke with them than with enclosed units. You will need to monitor the internal temperature and add wood chips or charcoal briquettes to maintain a proper temperature. For best use, fill the bottom of the grill with briquettes and burn them until gray ash appears. Separate the coals onto two sides of the grill and place a pan of water between them. Place the grate over the top and place the meat above the water. As the meat heats and cooks, the fat will drip into the heated water and create steam that will help destroy harmful bacteria. Keep the vents open on the cover. You will need to maintain an air temperature between 225°F to 300°F (107°C to 150°C) throughout the process.

*A barrel smoker* can be fashioned from a clean, uncontaminated 50-gallon (189 L) metal barrel with both ends removed. Set the open-ended barrel on the upper end of a shallow, sloping, covered trench or 10- to 12-foot (3 to 4 m) stovepipe. Dig a pit at the lower end for the fire. Smoke rises naturally, so having the fire lower than the barrel will aid its movement toward the meat. Mound the dirt around the edges of the barrel and fire pit to eliminate leaks. You can control the heat by covering it with a piece of sheet metal.

Use metal or wood tubes as racks for suspending hanging items, such as sausages, in the barrel. Metal strips can be attached to the cover to help hold it in place, trapping the smoke near the meat. You can monitor the inside temperature by suspending a thermometer from one of the racks. At the beginning of the smoking process, you'll want a rapid flow of air past the meat to drive off excess moisture. Less-rapid air movement near the end of the smoking period prevents excessive shrinkage of the meat. Once your fire is going, you can add green sawdust or green hardwood to cool the fire and make more smoke. *Never* use gasoline or other accelerants to start your fire. Besides their explosive potential, which can cause serious injury, the fumes and residues will contaminate your meat.

## POWERING YOUR SMOKER

A vertical water smoker is built with a bottom fire pan that holds wood chips or small briquettes and generally has two cooking racks near the top. The water pan positioned above the coals supplies moisture and helps regulate the internal temperature. An electric smoker is similarly constructed, except the smoke is controlled by premoistened wood chips rather than charcoal. This provides a more constant temperature and may require less attention during smoking. The sizes of electric smokers vary, with some accommodating several pounds (3 kg) of meat at one time.

Electric smokers are less likely to overheat and thus simply cook the meat because they can be more precisely controlled than any other type of smoker. If you're purchasing an electric smoker, make sure it has adjustable temperature controls that can reach 160°F (71°C) or higher.

### LIQUID SMOKE

If you're after a smoky flavor and not concerned about changing the texture or moisture level of your meat, you can try using liquid smoke. It can be added to a marinade or dry rub and will adhere to the surface of the meat before it's cooked. This coating will provide a smoky flavor. It should be used in moderation and in keeping with recipe recommendations to avoid off flavors.

Liquid smoke is produced by burning hardwood chips or sawdust such as hickory, apple, or mesquite at high temperatures. As the smoke passes through a condenser, it cools. When aided by water, this cool smoke will form a liquid. This liquid is then concentrated to create a stronger flavor. Liquid smoke is commercially available, and many supermarkets sell it as a flavoring and food preservative. It is widely used in foods in which a smoky flavor is expected, such as in bacon, smoked cheeses, tofu, and jerky.

Be aware that liquid smoke may contain some residual carcinogens because it is made from real smoke, although commercial production attempts to remove all smoke condensates such as tar and ash during processing. Some concern has been expressed relating to the different concentrations of polycyclic aromatic hydrocarbons (PAHs) that are found in different liquid smoke flavorings. These concentrations appear to relate to different types of trees used for the wood or sawdust, but they have been found to be below acceptable health levels. Manufacturers filter the liquid smoke in their production process and it's generally considered safe to use if used in moderation.

While wood or charcoal can be used for powering more traditional smokers, charcoal briquettes are difficult to use because they need to be kept at a constant and consistent burning rate for the ten hours required.

Note that when smoking, the smoke particles attach to the outer surface of the meat as the particles move or migrate from a warm surface (chamber) to a cold surface (meat). Even if you like a smoky flavor to your food, it is best not to overdo it. Too much smoke can make the meat taste bitter or like ash. Each smoker is different, and you may have to test several batches until you reach a flavor that suits you. Start light and work your way up incrementally if you don't want to ruin any meat. Keep records of your experimentation in a cooking log.

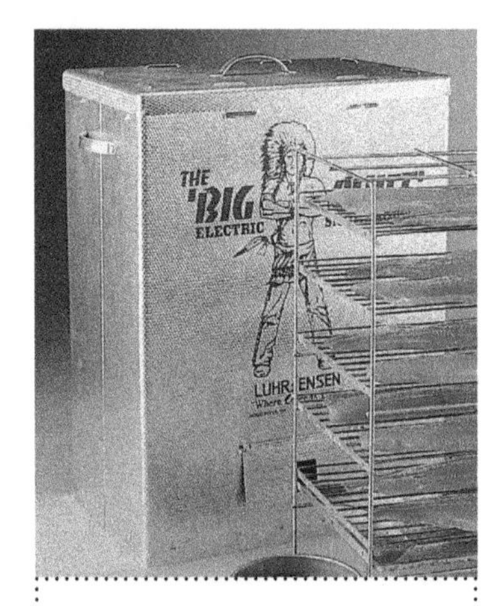


Electric smokers are popular because they are easy to use and reliable. There are stovetop smokers that can also be used.

A vertical smoker will have three sections. The meat to be smoked is placed at the top along with a thermometer to monitor the chamber temperature. The center has a pan with water to slow the smoking process and keep the meat from drying too quickly. The bottom is for coals, wood chips, or wood pieces. The smoke and heat rises to the top to cook the meat.

## WOOD TYPES FOR SMOKING

Selecting the type of wood to use for smoking meat is more a matter of personal preference than anything. Your choice will have an impact on the flavor and, often, the color. Combustion is created by heat produced by gas, electric, or pellets. When the heat is mixed with wood, it produces gases that create distinct flavors.

Different woods will create subtle, but different, flavors. Natural wood smoke is generally produced from hardwood sawdust, wood chips, or small logs. Woods can be divided into two basic groups that are based on whether they yield a mild or strong flavor, rather than tree species. For mild smoke flavors, use alder, apple, cherry, maple, orange, or peach woods. For stronger flavors, use hickory, oak, mesquite, pecan, and walnut.

One standard rule applies: the best woods for smoking are dried (cured) hardwoods with a low sap flow. Avoid using pine or other coniferous trees because of their high tar content, which will cause a bitter flavor.

## WOOD CONSIDERATIONS

Natural wood smoke contains three major components: solids, such as ash and tar; air and combustion gases; and acids, carbonyls, phenolics, and polycyclic aromatic hydrocarbons (PAHs).

Research has shown that the ash, tar, and gases do not contribute very much to the flavor, aroma, or preservative properties of smoked products. The phenolics have been identified as the primary source for flavor and aroma and the carbonyls are the source of color, typically the amber brown, generated from the smoking process.

If you use natural wood for smoking, it is important to use only air-dried woods and *never* use moldy woods that may contain toxins, woods that have paint on them, or

Most small home smoking units use wood chips to create the smoke while larger smoke houses can use split wood chunks to feed a fire. Chips are commercially available and can add different flavors depending on the wood type used.

woods that have been treated. Many of the woods you should use can be purchased at specialty stores, outdoor outlets, or even cut and dried yourself.

The dry wood used for smoking usually needs to be soaked in water first for about five minutes. This is because completely dry burning wood will create some smoke but not enough for your purposes. Moist wood, on the other hand, will smolder and create more smoke than dry wood. You want to create smoke, not a flame, to add as much flavor as possible. If using wood chips, you can soak them in water for at least an hour prior to using them. Small chips or sawdust can be sprayed with a mister bottle to dampen them instead of submerging them in water.

## SMOKEHOUSES

If you are considering an annual butchering schedule, you may want to consider constructing a stationary smokehouse for long-term use. While these are more elaborate structures than the smoking units previously discussed, they will accommodate larger quantities of meat at one time and will last for many years. They have the advantage of making temperature control easier, and their tight construction and well-fitted ventilators can precisely control airflow past the meat. Meat can be crowded into a smokehouse to some degree, but the rule is that no piece of meat touches another or the wall.

The purpose of a smokehouse is to enclose heat and smoke and reduce, but not entirely eliminate, airflow. Depending on how much smoking you want to accomplish, you can construct your own smokehouse or purchase a commercial unit. Because smokehouses are generally located outdoors, you should check if any local ordinances or fire codes apply before you begin construction of any new structure.

A home smokehouse is a simple version of a heat processing unit used by today's meat industry. The size may be vastly different, but the principles are the same: it is an enclosed area where the temperature and smoke level may be controlled with acceptable accuracy for meats, fowl, and fish. Smokehouses can be built out of cement or concrete blocks, and if you decide to build a smokehouse, it does not need to be elaborate to do an effective job. However, it must be adequately built and have the ability to be monitored so that the meat is properly cooked to minimize health risks. A sound smokehouse can last for years.

Other, perhaps unexpected, advantages of a smokehouse include a reduction in potential fire hazards (since the entire building will be built for smoking) as well as whole-building temperature control that can be dialed in even better than many smokers.

A larger-size building will provide space for several tiers of racks, allowing you to adjust the hangers to the size of the pieces of meat being smoked.

Always separate meat when smoking so pieces do not touch. Use equal-size pieces when cooking for the same length of time, as this will provide more uniform smoking results.

Smoking fresh sausages can be done safely when they are held at an appropriate internal temperature. Monitoring them throughout the process will help prevent them becoming overdone.

Materials used to build a smokehouse need to be able to withstand at least 160°F (71°C). They can be made of wood, concrete blocks, metal, or other sturdy, heat-tolerable materials.

Any smokehouse you construct should have four features: a source of smoke, a place to hold smoke, a way to hold the meat in the smoke, and a draft regulator near the top or bottom. Think of the smokehouse as a very slow oven in which the temperature does not exceed 200°F (95°C). Even though you will use and maintain low temperatures, build your smokehouse in a safe location away from other buildings, particularly your home, and away from combustible materials. You can get construction diagrams from many contractors or your county extension service.

The size of your smokehouse can be calculated based on the amount and weights of meat to be smoked. These requirements vary with the weight of the cuts. To estimate the capacity of your smokehouse, use an accepted measure of 12 inches (30 cm) in width, front and back, and 2 feet (61 cm) in height for each row that you plan to make use of.

Tiers used for hanging meat should start at least 18 inches (46 cm) above the floor, with each succeeding row 14 inches (35.5 cm) above the one just below it.

In order for the unit to work properly, air must be able to draft in from below and exit out the top. The degree to which this is controlled will determine the heat buildup and degree of smoke that remains in the unit. To create this air flow, make sure any unit you build has several holes near the top of each side, perhaps 2 inches (5 cm) in diameter. If the base of your smokehouse is tight to a foundation of cement, stone, or brick, drill two holes of the same diameter near the base of each side. You can further control draft by placing galvanized steel covers over the ventilation holes to be used as dampers. Small screens can be fastened to the holes on the inside of your smokehouse to keep pests and insects from entering it. Likely they will not approach while smoking is in progress, but there will be times when the unit is not in operation. You can also use small holes drilled into the sides of the smokehouse to insert thermometers to monitor the internal temperatures.

**NOTE:**
While smokehouses are excellent for processing meats, they do not make a good storage area for smoke-finished meats. After your smoking process is complete, flies will eventually get in.

Smoking units do not need to be elaborate, but they must be safe to use. They should have the ability to regulate the fire, heat, and ventilation. Above all, make sure that no fire can escape.

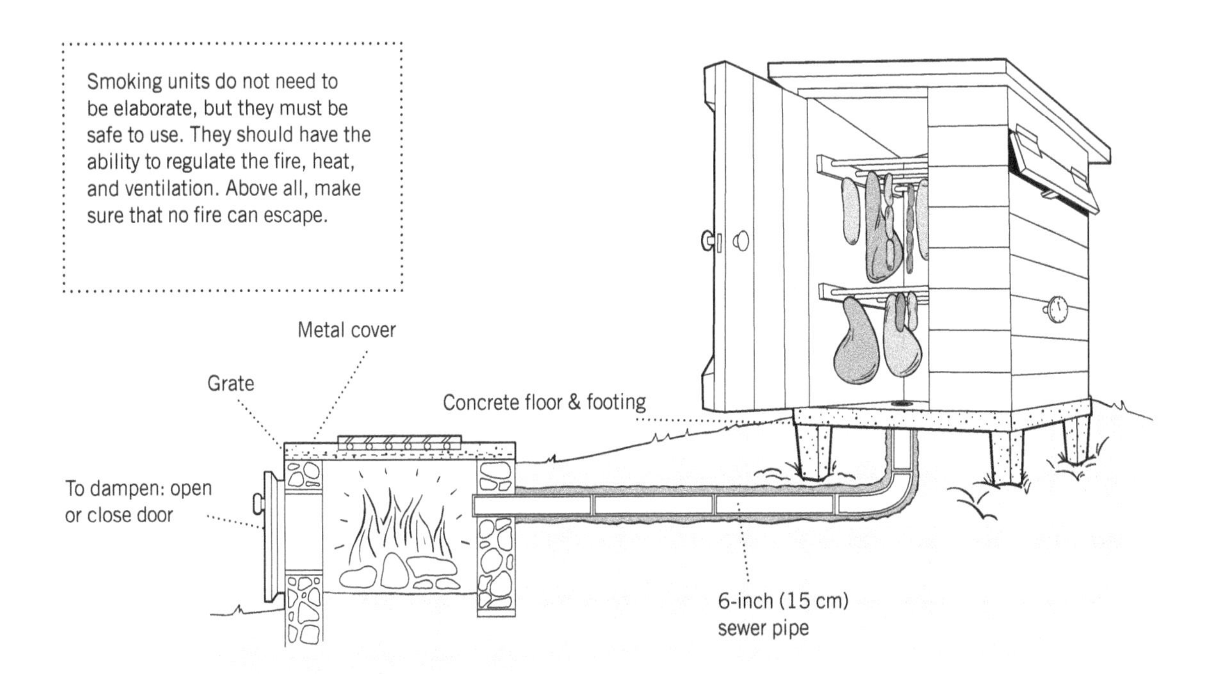


Any smokehouse you construct or purchase will need three basic components: a firebox for wood, a pipe to transfer the smoke from the fire to the meat, and a series of wood or metal tiers from which to suspend the meat.

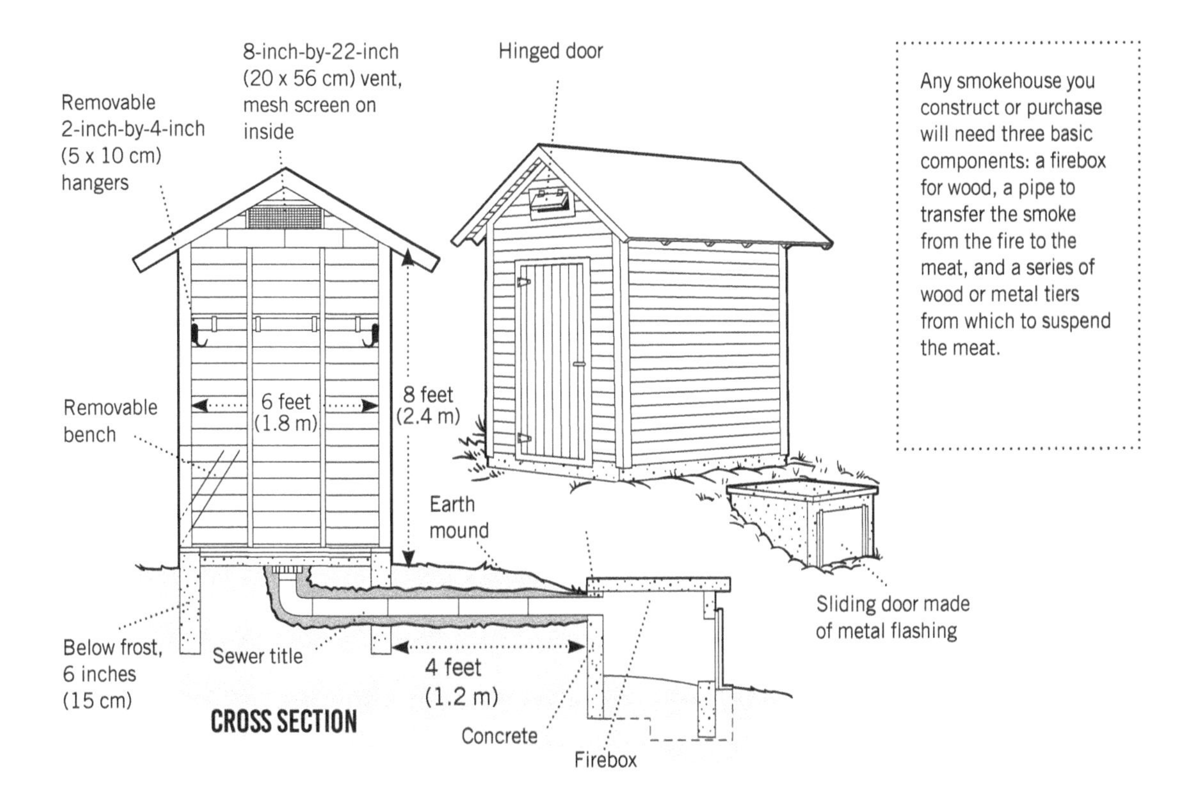

CHAPTER 10

# SAUSAGES

**SAUSAGE HAS BEEN A HIGHLY PRIZED MEAT PRODUCT THROUGHOUT MUCH OF MANKIND'S HISTORY AND IS ONE OF THE OLDEST KNOWN FORMS OF PROCESSED FOOD. THE HISTORY OF SAUSAGE PRODUCTION EXTENDS BACK TO THE ANCIENT BABYLONIANS, WHO PRODUCED AND CONSUMED SAUSAGES 3,500 YEARS AGO. EVEN HOMER'S *ODYSSEY*, WRITTEN IN THE EIGHTH CENTURY, BCE, REFERRED TO SAUSAGE.**

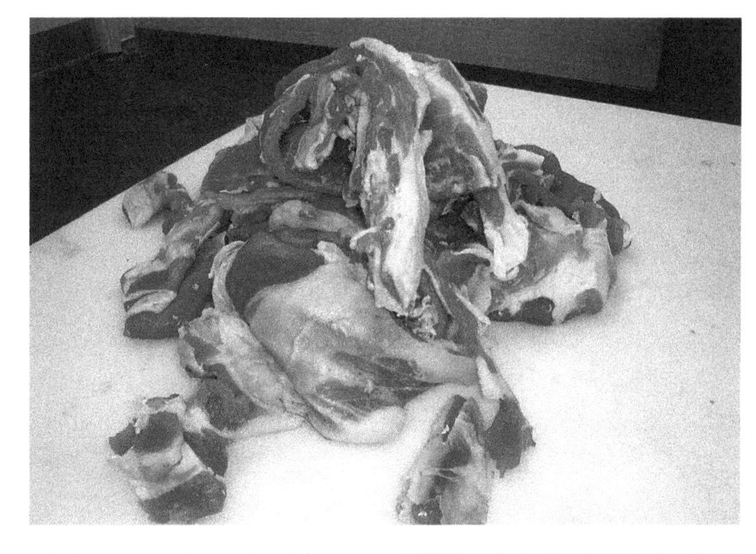

Europe saw the production of a variety of sausages by the Middle Ages. The cooler climates of Germany, Austria, and Denmark produced more fresh and cooked sausages because preservation was less of a problem. More temperate and warmer climates, such as in Italy, Spain, and southern France, developed dry or semidry sausages, which remain today.

The discovery of and subsequent demand for spices in Europe made them very important and valuable commodities and eventually allowed European sausage makers to become skilled at creating new and distinctive products bearing their influences. Some of these included types that are still well known today, such as bologna from Bologna, Italy, and braunschweiger from Brunswick, Germany.

Sausage making did not develop in the United States on an industrial scale until after the Civil War, although early Native Americans produced a type of sausage called pemmican. This was made by combining meat with dried berries and pressing it into a cake or a skin that was smoked or sundried.

The influx of European immigrants after the end of the Civil War brought many German, Polish, Italian, Dutch, Danish, and other nationalities with sausage-making skills to the United States. As these groups settled throughout the country, they brought their recipes with them, extending their influence and tastes wherever they went.

Many sausages consist of less valuable parts of animal carcasses, such as meat trimmings and fats that are less edible or might otherwise be discarded, which have accumulated from carcass fabrication. Fats add juiciness and flavor to sausages but add little nutritional value to the final product.

Today's sausage industry is diverse in size and type of production. Most major sausage processing plants in the United States are highly mechanized and automated to handle large volumes of products with speed and efficiency.

A niche production venue has been slowly developing with home sausage making. This chapter will provide information about basic types of sausages that can be produced at home and their characteristics; the safety and sanitation issues involved; spices, additives, and casings used; and the stuffings used in the process.

Homemade sausages are popular among hunters, who prefer to use all parts of the wild game they bring home. While home sausage making typically has been associated with rural areas, urban residents can shop in a market and make delicious and distinctive sausages with enough expertise.

## SAUSAGE VARIETIES

Of all the varieties of sausage produced in this country, the U.S. Department of Agriculture (USDA) only classifies them as two types: uncooked, which includes fresh bulk sausage, patties, links, and some smoked sausages; and ready-to-eat, such as dry, semidry, and cooked sausages.

Like other fresh meat, fresh sausages are highly perishable and must be refrigerated or frozen until ready to be cooked. Fresh sausages must be cooked prior to consumption to avoid health risks. Ready-to-eat sausages have been processed and preserved with salt and spices and may be dried or smoked. These types of sausages, such as jerky or sticks, can be eaten out of hand or cooked and heated, like hot dogs.

## FRESH SAUSAGES

Fresh sausages are made from uncooked and uncured cuts of meat. These sausages include those seasoned and stuffed into casings, or those in bulk form that will be pressed into patties. However, they are not cured or smoked. Fresh sausages should be eaten within three days of processing or purchase (or frozen), and they should be thoroughly cooked before being served. Examples include the following:

*Bockwurst*: A German-style sausage made from ground veal or veal and pork combined. It is typically flavored with onions, parsley, white pepper, paprika, or cloves and often sold fresh. Cooked bockwurst has an increased shelf life and is usually cooked by simmering, although it can be grilled.

*Bratwurst*: A German-style sausage made from pork, beef, and veal. It looks like a big hot dog and is flavored with allspice, caraway, and marjoram, although recipes can vary between regions and countries. It can be produced as fresh or cooked.

*Chorizo*: Originating in Spain, the term encompasses several types of pork sausage. It can be fresh or cured. Fresh chorizo is similar to Sicilian sausage but is much spicier. Cured or dried chorizo can resemble pepperoni in size and shape but has a sharper taste and smell. Different countries have different recipes for making chorizo, and some have a sweet or spicy flavor.

*Country-style or breakfast sausage*: One of the most common kinds of sausage found in the United States. It is known by several names, can be made into patties or small links, and is flavored with sage, savory, and thyme.

*Pork sausage*: A fresh, uncooked sausage made entirely from pork and seasoned with salt, pepper, and sage. It is often sold in bulk, in a chub or link form, or as patties.

*Kielbasa*: Similar to Italian sausage in that its name is more of a generic term than a reference to a specific sausage. In the United States, it refers to a Polish or Polish-style sausage. It is typically made from coarsely ground lean pork and is sometimes combined

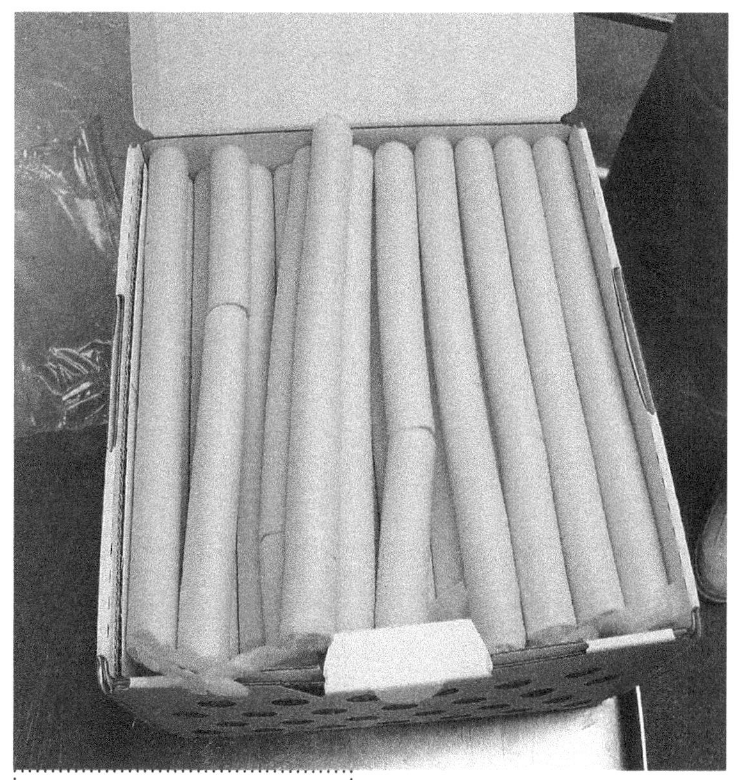

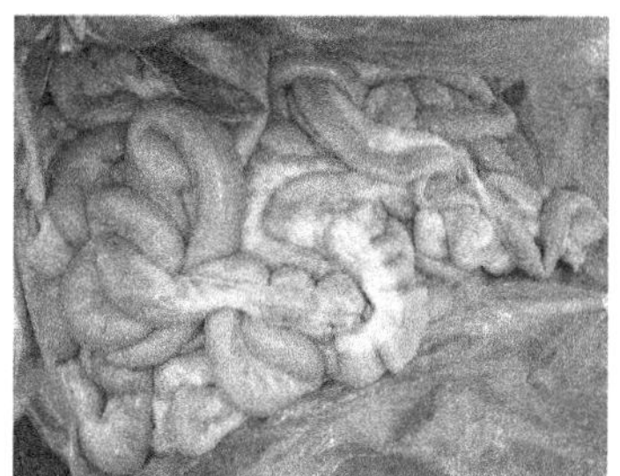

Natural casings are made from animal intestines, particularly sheep, pork, and beef. When using intestines for casings, they must be thoroughly cleaned and washed. The intestinal membranes are strong, flexible, and resilient, but they tend to lack the uniformity of manufactured casings.

Thin, light-colored casings manufactured from beef collagen are often used for fresh sausages. Certain types cannot be used for smoking because they can dissolve during the cooking phase and the sausage will fall out.

Cut the intestines into 2- or 3-foot (61 to 91 cm) sections and strip the fecal content from them. Flush with clean water and then soak in a salt brine to neutralize the remaining contaminants. Keep immersed until used because if they dry out, they will crack and break when stuffed with sausage. Natural sausage casings should be packed in salt for storage.

Artificial or manufactured casings come in several sizes, thicknesses, and color for different types of sausages. Most manufactured casings must be soaked in clean water before use to make them pliable.

with beef or veal, or both. Commercial kielbasa is usually an uncooked, smoked sausage with a medium red color.

*Italian-style sausage*: A fresh sausage that must be fully cooked before eating and can have either a hot or sweet taste. It is traditionally a pure pork sausage with pepper, fennel, and other spices as the added ingredients for flavoring.

*Liverwurst*: A popular German-style sausage. It is made from finely ground pork and pork liver. It can be stuffed into a nonedible casing but must be thoroughly cooked before being served. Spices such as ground black pepper, marjoram, allspice, thyme, ground mustard seed, and nutmeg are used to provide distinctive flavors. The term is sometimes interchanged with Braunschweiger because of the similarities between the two in production, taste, and texture.

*Thuringer sausage*: A lightly smoked, German-style sausage similar to summer sausage. It is often semidry and is more perishable than other cured sausages, even though, technically, it is cured. Some are not fermented and are sold fresh. It is mostly made from pork, but beef and sometimes veal can be used. Flavorings that are used to make Thuringer sausage are used in fresh pork breakfast sausages but without the sage.

*Haggis*: A Scottish traditional food made from the heart, lungs, and livers of sheep or calves. It is highly seasoned, mixed with oatmeal, onions, suet, spices and salt, and then stuffed into a sheep's or pig's stomach. It is then boiled for about three hours. Haggis is perhaps the only sausage that is involved with the sport of hurling it for distance, called haggis hurling.

## COOKED AND SMOKED SAUSAGES

Cooked sausages are usually made from fresh meats that are cured during processing, fully cooked, and/or smoked. Cooked sausages should be refrigerated until eaten. They will generally keep seven days after being opened. Because they are fully cooked, they are ready to eat once opened, although you may prefer to serve them warm or hot. Examples include the following:

*Frankfurters*: Also known as the common hot dog, these are touted as the most consumed sausage in the world. Processed hot dogs contain mostly water and fat and have a soft, even texture and flavor. Homemade frankfurters can be made with a blend of beef, pork, and/or poultry meat. In the United States, if fillers are used, such as cereal or soy, the name must be changed to "links" or their addition must be identified on any sales label.

*Bologna*: A generic term for a fully cooked, mildly seasoned sausage made from low-value pieces of beef, pork, or both. It can be eaten cold or reheated. Bologna is usually produced in large-diameter rings or chubs, which give it several distinctive styles and shapes although they are constitutionally much the same as hot dogs. Beef bologna is an all-beef version and will appear a more red color because it does not have a mixed meat composition. It can also be made from pork, turkey, or chicken.

*Vienna sausage*: Sometimes called garlic sausage, it is made in the general shape of a hot dog, although it can be longer and somewhat thinner. It is a sausage with a creamy meat texture and is made primarily from pork and beef, although chicken and turkey can be used. Veal is sometimes added to create a milder flavor. The predominant flavors include onions, mace, and coriander. Sometimes pistachio nuts are added for seasoning.

*Beerwurst or bierwurst*: A large sausage, usually 2 to 3 inches (5 to 7.5 cm) in diameter or larger, of a dark red color. It is stuffed into veined natural casings or vein-decorated artificial casings. It is made from coarse-ground beef and pork and spiced with garlic, black peppercorns, paprika, and mustard seeds. Contrary to its name, it does not contain any beer. It is usually sold as sandwich meat.

*New England sausage*: Also known as Berliner, this sausage is made from coarse-ground pork with pieces of ham or chopped beef interspersed within it. Generally, it is stuffed into large casings, similar to beerwurst.

*Braunschweiger*: A creamy-textured, German-style liver sausage of pure pork origins. It has a mild flavor that includes onions, mustard seed, and marjoram. It is nearly always smoked and generally served cold as a spread for toast or used as a filling for sandwiches.

*Mettwurst*: A strongly flavored German-style sausage made from raw minced pork and preserved by smoking and curing. It contains ginger, celery seed, and allspice. Although it is smoked, it needs to be cooked thoroughly before being served. Mettwurst can have either a soft or hard texture, depending on the length of smoking time used.

# DRY AND SEMIDRY SAUSAGES

Dry and semidry sausages are made from fresh meats that are ground, seasoned, and cured during processing. They are stuffed into either natural or synthetic casings, fermented, often smoked, and carefully air-dried. True dry sausages are generally not cooked and may require long drying periods of between twenty-one to ninety days, depending on their diameter.

The distinctive flavor of these sausages is due to the lactic acid produced by fermentation. This fermentation occurs after the meat is stuffed into casing and the bacteria metabolize the sugars, producing acids and other compounds as byproducts and the resulting tangy flavor.

Semidry sausages, such as summer sausage, are often fermented and cooked in a smokehouse. Both dry and semidry sausages are ready to eat and do not require heating before serving, although a cool temperature or refrigeration is recommended for storage. Dry and semidry sausages include summer sausage, pepperoni, salami, and Landjäger, among others. These are detailed in the paragraphs below.

*Summer sausage*: A general term for any sausage that can be kept without the need of refrigeration. It is typically a fermented sausage with a low pH to slow bacterial growth and provide a longer shelf life. It is usually made from a mixture of beef or beef and pork. Venison can also be used to make summer sausage. It resembles some of the drier salamis but is milder and sweeter in flavor. Summer sausage can be either dried or smoked, and although curing agents can vary considerably, some sort of curing salt is almost always used.

*Pepperoni*: A hotly spiced Italian-style sausage made from coarse-ground, fermented pork with ground red pepper as the main flavor ingredient. It is a dry sausage and increases in flavor as it progresses through the drying process.

*Salami*: Not necessarily a specific sausage but most often refers to those products that have similar characteristics and is made from beef, pork, or both. Salamis can be found in many sizes and shapes, and they may be dry and quite hard. Most are made with garlic, salt, various herbs and spices, and some minced fat. Salamis are made by allowing the raw meat mixture to ferment for twenty-four hours before it is stuffed into either a natural or synthetic casing and then hung to dry. Most are treated with an edible mold culture that is spread over the outside, which prevents spoilage during curing. Pepperoni is one type of salami, and others include Genoa, kosher, Milano, Sicilian, Novaro, and Sorrento.

*Landjäger*: A traditional Swiss-German dried sausage that is a popular snack food. Its taste is similar to dried salami, and it can be boiled and served with vegetables. It is made from equal portions of beef and pork, with fat or lard and sugar and spices added. The meat is pressed into small casings for making links, usually 6 to 8 inch (15 to 20.5 cm) lengths. They are then pressed into a mold before drying. This gives the strips their characteristic rectangular shape. After drying, they can keep without refrigeration if needed.

## SPECIALTY SAUSAGES

This group of sausages may include cured, uncured, smoked, and nonsmoked meats that do not readily fit into other groups. These include luncheon meats that may be cooked in loaf pans or casings, or water cooked in stainless-steel molds. Some specialty sausages include headcheese, olive loaf, scrapple, and souse.

*Headcheese*: Actually not made from cheese but, rather, from meat pieces from the head of a calf, pig, sheep, or cow set in a gelatin base. It may also include meat from the feet, tongue, and heart. It originated in Europe and is usually eaten cold or at room temperature as a luncheon meat. In the United States, headcheese is available in two styles: French and German. French headcheese is made from cured pork pieces suspended in a vinegar-flavored gelatin with small bits of pickles and pimentos added for flavor. There is no casing. The German style, also composed of cured pork and beef pieces, differs in that the gelatin is produced by heating to resemble an emulsion and, because of this, requires a casing.

*Olive loaf*: In reference to meat, olive loaf refers to a type of bologna that is composed of cured pork and beef, seasonings, sweet pickles, diced pimentos, and green olives. It is served as a luncheon meat and can also contain garlic, basil, and sweet peppers.

*Scrapple*: Not a sausage per se, but rather a cornmeal-based mixture or mush of head meat, pork sausage, or both. It can also contain pork scraps and trimmings that can't be used elsewhere. Meat scraps that come from the head, heart, liver, or attached to bones are boiled to make a broth. When they are finished cooking, the bones and fat are discarded, and the meat is then mixed with dry cornmeal and added to the broth, which is then boiled. Seasonings such as sage, thyme, savory, and black pepper are added. This mush is formed into a semisolid congealed loaf and allowed to cool until thoroughly set. Then, slices are pan fried before serving. It has a Pennsylvania Dutch origin with strong regional recognition, although it can be found throughout the country.

*Souse*: A loaf product similar to headcheese. It typically contains cured or cooked meat from the heads, tongues, lips, and snouts of pigs, which are added to gelatin. Pickles and pimentos are added for color and flavor.

## VENISON SAUSAGE

Venison sausage is made from deer or other large game animals. Because of these animals' feeding habitat, their meat tends to be very dry. Pork and pork fat are typically added to enhance venison's flavor and palatability. It may be prepared as fresh, dried, or smoked sausage in ways similar to domestic animal-based sausages.

# OTHER SAUSAGES

It is estimated that there are more than two hundred types of sausage and related products produced today, with many variations of each. While you may start with producing only one or two types of sausage, in time you may venture to others. The list may include the following:

*Alesandri*: An Italian-style member of the salami family made with highly seasoned cured pork

*Arles*: A French-style salami that contains coarsely chopped pork and beef seasoned with garlic

*Berliner*: A pork and beef sausage mildly flavored with salt and sugar

*Blood and tongue sausage*: Contains cooked lamb and pork tongues and hog blood

*Bloodwurst or blood sausage*: Sausage made of pig blood, pork meat, ham fat, gelatin-based meats, salt, pepper, clover, allspice, and onions

*Boterhamwurst*: A Dutch-style veal and pork sausage

*Corned beef*: Made from cured and spiced beef brisket

*Cotto salami*: Cooked salami enhanced by whole peppercorns

*Easter nola*: An Italian-style mildly seasoned, salami-type dry sausage

*Garlic sausage*: Similar to the frankfurter in taste and texture but with a more pronounced garlic flavor

*Farmer sausage*: Originated with farmers of northern Europe and is a mildly seasoned dry or semidry sausage made of 65 percent beef and 35 percent pork. Chopped medium fine, seasoned, stuffed into beef middles casings, and heavily smoked.

*German salami*: Similar to the Italian-type salamis but more heavily smoked

*Goteborg*: A Swedish-style cervelat that is heavily smoked and made from coarsely chopped pork and beef that is flavored with cardamom

*Ham and cheese loaf*: A specialty product containing pieces of cheese imbedded within finely ground ham

*Holsteiner*: Similar to farmer sausage, except that the ends are tied together like a horseshoe. Dried and smoked.

*Honey loaf*: A mixture of pork, beef, honey, and spices but can also contain pickles or pimentos, or both

*Hungarian salami*: A mild, dry salami made of lean pork and backfat

*Knackwurst*: Similar to the frankfurter in texture and mixture and is fully cooked but with more garlic added

*Linguisa*: A Portuguese pork sausage cured in brine, seasoned with garlic, and spiced with cinnamon and cumin

*Liver loaf*: A sandwich-shaped liver sausage that is similar in flavor to liverwurst

*Longaniza*: A dry Portuguese sausage flavored similarly like chorizo

*Macaroni and cheese loaf*: Contains chunks of cheddar cheese and pieces of macaroni mixed with ground beef and pork

*Mortadella*: A dry sausage containing pork, beef, and cubes of pork fat, seasoned with anise

*Pastrami*: The cured, smoked plate of beef that is usually thinly sliced for sandwiches

*Pinkel*: A sausage made of beef, oats, and pork fat

*Salsiccia*: A fresh Italian sausage made of finely ground pork

*Smoky links*: Smoked, cooked links made from pork and beef spiced with pepper

*Straussburg*: A liver and veal sausage that contains pistachio nuts

*Vienna sausages*: The small hors d'oeuvre or cocktail-style frankfurters or hot dogs

*Weiswurst*: A fresh German-style sausage that is mildly spicy and made of pork and veal

# SELECTING INGREDIENTS, ADDITIVES, AND SPICES

The ingredients, additives, and spices you add, and their quality, will greatly affect the taste and texture of your final sausage products. For sausage making, ingredients you add may be raw meat and nonmeat materials. The interaction between the ingredients and meat materials used will determine the different flavors and textures between sausages.

You control the ingredients and spices you add to your sausage. By understanding the properties of each, you will be able to create products that meet your requirements and tastes, whether it is to limit the preservatives you eat or to avoid high fat products routinely available at retail markets.

You should keep one simple rule in mind when creating your sausages, regardless of which kind they are: your finished product is only as good as the ingredients it contains. The meat you start with should be fresh, have a proper lean-to-fat ratio, and exhibit good binding qualities. Clean meat that has been cut in sanitary conditions is a prime requirement. The meat used should not have been contaminated with bacteria or other microorganisms at any stage of processing or cutting.

## INGREDIENTS

Most of the ingredients you will use are readily available for purchase at local supermarkets or meat markets, or from other specialized commercial businesses. Licensed retail outlets that specialize in such products are another source. Internet businesses that sell ingredients for meat processing can provide them for sausage making. If buying ingredients from businesses outside your area, be sure to check their licenses and ask about the sources they access for their products. Always check their labels to be sure you know what you are adding to your meat.

The main ingredients for your sausage making are likely to be based on meat derived from pork, beef, and perhaps veal, either separately or in a combination. Other ingredients may include hearts, tongues, livers, kidneys, and stomachs. If you don't raise your own livestock for home use, be sure the meat you buy comes from a reputable source or that it has been USDA inspected. If using your own animals for processing, be certain they are healthy and disease free. Cuts with the lowest economic value are generally used for sausage making.

Several nonmeat ingredients are used to provide flavor, inhibit bacterial growth, and increase the amount of sausage produced. These may include water, salt, sugar, nonfat dry milk, soy products, extenders and binders, and spices.

## BINDERS AND EXTENDERS

Several ingredients are often added during commercial sausage production that are referred to as binders and extenders. Some of these products are used for both purposes. Binders are used to help the meat particles adhere to each other or to prevent them from separating during the production process. Extenders are used to increase the moisture content and texture of the product, as well as stretch the amount of product derived from a certain volume of meat.

Extenders often include non-fat dry milk and similar dried products of milk origin including dried whey. Binders can include many derivatives of milk plus cereal flours, wheat gluten, and soy flours.

Other components of sausage production that are often added are:

*Water and ice*: Added sometimes to add moisture and keep the sausage cold during processing. Cold temperatures delay bacteria growth and add to the final product quality. Water also helps dissolve salts for better distribution within the meat.

*Salt*: Serves three functions in the meat: preservation, flavor enhancement, and draws out protein to help bind the mixture. Sodium nitrate and nitrite are used for curing meat as they inhibit growth of a number of pathogens and bacteria that cause spoilage, including those that cause botulism. Nitrate and nitrite are the most regulated and controversial of all the sausage ingredients. It is strongly recommended that a commercial premixed cure be used when nitrate and/or nitrite is called for in the mixture.

*Sugar*: Used for flavor and to counter the bitter taste of salt. It helps reduce the pH in meat because of the fermentation of the lactic acid.

*Ascorbates and erythorbates*: Vitamin C derivatives that speed the curing reaction. They can be used interchangeably in cured sausages to which nitrite has been added.

## SPICES, SEASONINGS, AND FLAVORINGS

Many different spices, seasonings, and flavorings are used in sausage production to increase taste. For home sausage making, they are generally added by personal preference and taste or to follow general guidelines for a particular product recipe. By combining different levels of various spices, you can create unique and distinctive sausages.

Spices, seasonings, and flavorings are not usually included to add to the nutritional value of the sausage, although some minute traces of nutrition are provided by them. Spices vary greatly in composition and may be added as whole seeds, coarsely ground, or in powdered form. Some of the major spices used include:

*Allspice*: A reddish brown pimento berry sold whole or ground. Pungent, clove-like odor and taste. Used in bologna, pork sausage, frankfurters, hamburgers, potato sausage, headcheese, and other products.

*Basil*: Marketed as small bits of green leaves, whole or ground. Aromatic, mildly pungent odor and used in dry sausage such as pepperoni.

*Bay leaves*: Elliptical leaves marketed whole or ground. Fragrant, sweetly aromatic with slightly bitter taste. Used in pickling spice for corned beef, beef, lamb, pork tongues, and pigs' feet.

*Caraway seed*: Curved, tapered, brown seeds sold whole. Slightly sharp taste and used in Polish sausage.

*Cardamom seed*: Small, reddish brown seeds sold whole or ground. Pleasant, fragrant odor; used in bologna, frankfurters, and similar meats.

*Cloves*: Reddish brown, sold whole or ground. Strong, pungent, sweet odor and taste. Used in bologna, frankfurters, headcheese, liver sausage, corned beef, and pastrami. Whole cloves can be inserted into hams and other meats during cooking.

*Coriander seed*: Yellowish brown, nearly globular seeds sold whole or ground. Lemon-like taste. Used in frankfurters, bologna, knackwurst, Polish sausage, and other cooked sausages.

*Cumin seed*: Yellowish brown oval seeds sold whole or ground. Strong, bitter taste; used in chorizo and other Mexican and Italian sausages; used in making curry powder.

*Dill seed*: Light brown oval seeds sold whole or ground. Warm, clean, aromatic odor used in headcheese, souse, jellied tongue loaf, and similar products.

*Garlic, dried*: White color ranging in forms of powdered, granulated, ground, minced, chopped, and sliced. Strong, characteristic odor with pungent taste. Used in most beef sausages and salamis.

*Ginger*: Irregularly shaped pieces brownish to buff-colored; sold whole, ground, or cracked. Pungent, spicy-sweet odor; clean, hot taste. Used in pork sausage, frankfurters, knackwurst, and other cooked sausages.

*Mace*: Flat, brittle pieces of lacy, yellow to brownish orange material sold whole or ground. Somewhat stronger than nutmeg in odor and flavor. Used in bologna, mortadella, bratwurst, bockwurst, and other fresh and cooked sausages.

*Marjoram*: Marketed as small pieces of grayish green leaves either whole or ground. Warm, aromatic, slightly bitter and used in Braunschweiger, liverwurst, headcheese, and Polish sausage.

*Mustard*: Tiny, smooth, yellowish or reddish brown seeds sold whole or ground. Used in bologna, frankfurters, salamis, summer sausage, and similar meat products.

*Nutmeg*: Large, brown, ovular seeds sold whole or ground. Sweet taste and odor. Used in frankfurters, knackwurst, minced ham sausages, liver sausage, and headcheese.

*Onion, dried*: Similar to garlic. Used in luncheon loaves, Braunschweiger, liver sausage, headcheese, and other meat products.

*Oregano*: Marketed as small pieces of green leaves, whole or ground. Strong, pleasant odor and taste. Used in most Mexican and Spanish sausages, fresh Italian sausage, and sometimes in frankfurters and bologna.

*Paprika*: Powder form ranging in color from bright red to brick red. Slightly sweet odor and taste. Used in frankfurters, fresh Italian sausage, bologna, and many other cooked and smoked sausage products.

*Pepper*: Black, red, white in color and sold whole or ground. Penetrating odor and taste, ranging from mild to intensely pungent. Black pepper is the most used of all spices, but white is substituted when black specks are not wanted, such as in pork sausage and deviled ham. Red pepper is used in chorizo, smoked country sausage, Italian sausage, pepperoni, fresh pork sausage, and many other meat products.

*Rosemary*: Needle-like green leaves available whole or ground. Fresh, aromatic odor, somewhat like sage in taste. Used in chicken stews and other poultry products.

*Saffron*: Orange and yellowish in color, sold whole or ground. Strong odor and bitter taste. Most expensive of all spices and used primarily for color in a few sausages.

*Sage*: Grayish-green leaves sold whole, ground, or cut. Highly aromatic with strong, slightly bitter taste. Used in pork, pizza, and breakfast sausages.

*Savory*: Sold as dried bits of greenish brown leaves. Fragrant, aromatic odor and used primarily in pork sausage but also in other sausages.

*Thyme*: Marketed as gray to greenish brown leaves, whole or ground. Fragrant, odor with pungent taste. Used in pork sausage, liver sausage, headcheese, and bockwurst.

Whether they are added by volume or weight, herbs and spices are a very small percentage of any sausage but have an enormous influence on character and flavor of the end product. In either case, the best herbs and spices are those that are purchased fresh or homegrown and those used soon after harvesting. If you purchase either or both, try to buy new products rather than older ones because new vintages will have retained their potency more than older ones. Store herbs and spices in a cool, dry area away from heat and light. Freshly dried herbs and spices rarely retain their optimum flavors longer than six months.

## SALT AND PEPPER

Salt and pepper add flavor and aroma to sausages. Many different types of salt are available, but those without additives, such as iodine, provide maximum flavor. Pepper can be purchased as whole peppercorn and ground when needed. Recipes may make distinctions between three forms of pepper: finely ground, medium grind, and coarse grind. Finely ground is a fine powder with no large pieces in it. Medium grind refers to flakes that will pass through a typically shaker. Coarse grind has small bits that may be ground in either a pepper grinder with a coarse setting or with a mortar and pestle.

# NITRITE AND NITRATE CONCERNS

Curing meat with products containing nitrates and nitrites was previously discussed on pages 143 to 146. Please read those pages and make sure you have a thorough understanding of cures containing nitrates and nitrites as well as how they work before moving forward.

In short, you must learn about these ingredients before you use them as they can cause illness or even death if used improperly. Accidentally licking a finger covered with curing salt or leaving these ingredients within the reach of children could have dire consequences! That said, when used properly, you do not need to be scared of nitrates/nitrites in much smaller concentrations. They have an important function when making some types of sausages.

As previously discussed, concerns about consuming nitrates or nitrites in meat center on the quantity eaten rather than their inclusion as preservatives. Still, you can certainly focus on only fresh sausages if you are concerned about using nitrites and nitrates.

# CASINGS

It is not necessary to stuff fresh sausage meat into a casing. It can be left in bulk form or made into patties. But if ground into bulk form, it will have to be used within one or two days to retain its freshness and quality. Most sausages are made by inserting the ground ingredients into some forming material that gives them shape and size and holds the meat together for cooking and smoking, or both. This material is called a casing.

There are two types of casings used in sausage making: natural and manufactured. Although their purposes are the same, their origin is very different.

Natural sausage casings are made from parts of the alimentary canal of various animals that can include the intestinal tracts from pigs, cows, or sheep. One advantage for using them is that they are made up largely of collagen, a fibrous protein, whose unique characteristic is variable permeability. This allows smoke and heat to penetrate during the curing process but without contributing undesirable flavors to the meat. Natural casings can be purchased from companies that offer sausage-making products or they can come from an animal that you are butchering. Packing houses that save casings will flush them with water and pack them in salt before selling them to casing processors. The casing processor does the final cleaning, scraping, sorting, grading, and salting before you purchase them. If using your own animal casings, it is important they are thoroughly flushed and cleaned and are placed in a salt brine prior to use.

## CLEANING CASINGS FOR HOME USE

You can clean your own hog and sheep casings for sausage production after they are removed from the body cavity. Because they are unlikely to be the first parts you work with from the carcass, they need to be set in cold water to reduce their temperature to prevent spoilage. If working alone, you should set up the cold water tub prior to beginning the butchering process. If working with others, you can designate another to handle this part.

There are several things to consider if using the intestines for meat casings. The first is sanitation. The intestines will likely be filled with excrement that contains *E. coli* bacteria and needs to be kept away from any organs that you plan to use later. If you plan to use several or all of the organs attached within the viscera, you will need to cut them away from the mass of intestines and stomach prior to placing them into the cold water to reduce the chances of contamination. Cut the heart, liver, spleen, kidneys, or any other organs away from the intestines and stomach and place them aside in a clean container. The stomach and intestines should then be separated by making a cut at the point where the stomach and large intestines meet. If you have used a cord or string to tie off the end of the intestine at the anus when you made your cuts to remove it, then one end should still be tied. After making your cut at the stomach and intestine junction, place that end in an empty pail or other container and allow the intestinal materials to drop into it. Depending on the species and size of the animal you butchered, there may be much intestine to work with or relatively little. You may have to use your hand to strip as much of the excreta out of the intestine as possible. When finished, you can place the intestine in a cold water bath primed with salt and work with it later after you deal with the more valuable meat cuts of the carcass.

Intestines make very good natural casings for your sausages because they are largely collagen and will easily break down during the curing process, yet they still are strong enough to hold the meat during the stuffing process. Their flexibility makes them an attractive alternative to synthetic casings. But good casings are also clean casings and you will need to prep them for use by removing all of the excrement and intestinal linings before using them.

To begin properly cleaning your casings, you will need to invert the intestine by turning it inside out. After removing any cord or string that has closed one end, start by turning one end of the intestine inside out to create a lip much as you would roll down your socks, except that you are not making a rolled up mass. You want to pass the rest of the intestine through this roll as if you were peeling a banana without breaking it. When the inside of the intestine has completely become the outside, you can thoroughly wash it in a cold 0.5 percent chlorine solution. Use a soft-bristle brush to very gently scrub the excess fat, connective tissue, and any residual foreign or fecal materials off the intestine. Although the intestine can withstand some good scrubbing, you need to be careful not to overdo it or you may leave little tears in the membrane that can rupture and break during the stuffing process.

After thoroughly cleaning the intestine, rinse it with clean, cold water and invert it back to its original form. Use a saturated salt solution (1 gram salt/2.8 ml water) for storage overnight. If you are not using the casings for several days, they can be kept in this solution with cold temperatures. If not using them for two weeks or longer, you can freeze them in this salt-saturated solution. This solution will also inhibit the growth of bacteria that thrive on salt, as well as other bacteria that may have survived the chlorine solution bath. Before using your stored casings, gently flush them in lukewarm water to remove any clinging salt. They are then ready for use.

Generally, the larger the animal butchered, the larger the size of the intestinal tract you will have. The kind of sausage you want to make may have some bearing on the size of casings you use. Typically lambs and sheep are the smallest, followed by pigs, and then cows. Sheep casings are more delicate and can be used for hot dogs, frankfurters, and pork breakfast sausages.

Hog casings have a wide variety of use and are considered to be an all-purpose casing. They have five classes: bungs, middles, smalls, stomachs, and bladders. Bungs and middles are best used for dry sausage; smalls for fresh sausage, such as chorizos, bratwurst, bockwurst, and Polish; stomachs for headcheese; and bladders for minced luncheon meats. Small hog casings, from the small intestine, are probably the most widely used and easiest to find at a local meat shop.

Beef casings are larger and are classed as bungs, rounds, and middles. Rounds are the most common of all beef casings and come from the small intestine. They are used for ring bologna, Holsteiner, and mettwurst. Beef bungs are large casings that come from the cecum or appendix and may have diameters of 3 to 5 inches (7.5 to 13 cm), making them good for stuffing bologna. They can contain from 12 to 20 pounds (5.5 to 9 kg) of sausage.

## MANUFACTURED AND ARTIFICIAL CASINGS

The alternative to natural casings is a group of manufactured or artificial casings that are made from edible or inedible materials. Fibrous casings are popular because they are uniform in size and easy to use. They are made from a special paper pulp mixed with cellulose, are inedible, and must be peeled away before eating. However, they provide the most strength of any casing available. Three of the most common types of manufactured casings include collagen, cellulose, and artificial.

Collagen casings are made from the gelatinous substances found in animal connective tissue, bones, and cartilage and mechanically formed into casings. Because of their lower structural strength, these casings generally are made into small diameter products and are ideal for breakfast links or fresh, smoked, and dried sausages. Unlike large cellulose and fibrous casings, collagen casings should not be soaked in water before use. They are easier to work with when dry.

Cellulose casings are made from cotton linters, the fuzz from cotton seeds, which are dissolved and reformed into casings. Cellulose casings are crimped into short strands, and an 8-inch (20.5 cm) length may stuff as much as 100 feet (30.5 m) of sausage. Small cellulose casings work well for skinless wieners and other small-diameter skinless products.

Artificial casings are frequently made from plastic and are inedible. They can be used for sausages cooked in water or steam, such as bologna and Braunschweiger.

If you are not using your own casings, you can purchase any of the manufactured or artificial casings from meat-packing companies, sausage supply businesses, local butcher shops, or through ethnic markets.

# ESSENTIAL EQUIPMENT

You only need a few pieces of equipment to make sausage in your home whether or not you butcher your own animal. The three most important pieces of equipment you will need include a meat grinder (but not if you only purchase bulk sausage meat), a sausage stuffer, and a thermometer. Other pieces that you may find useful include a mixing tub, a scale, and a smoker if you want to do your own meat smoking and preservation. Sausage-making equipment is usually available from meat equipment supply companies.

## MEAT GRINDERS

A meat grinder is used to reduce the size of the meat pieces into a pliable mixture. They can be operated by hand or by electricity. Some food processors can do a good job of chopping meat, and some heavy-duty mixers may have a grinding attachment that will work.

Hand grinders have been used for generations and usually have several different-sized grinding plates or chopping disks, ranging from fine, with holes 1/8 inch (3 mm) in diameter, to coarse, with holes 3/8 inch (1 cm) in diameter. All hand grinders will have a screw augur that is attached to the outside handle. A disk cap screws over the top of the grinding plate to hold it in place while the meat is forced through the holes by the auger. It is a simple process once you have it set up. Hand grinders typically have a tightening screw at their base so that they can be mounted to a table or sturdy support frame.

If you are making small amounts of sausage, a hand grinder should be sufficient. Some large grinders make use of a small motor with belt attached. This is a fast, efficient way to grind a very large amount of meat in a very short amount of time. Food processors can be useful in producing finely ground or emulsified sausages, such as frankfurters, bologna, and some loaf products.

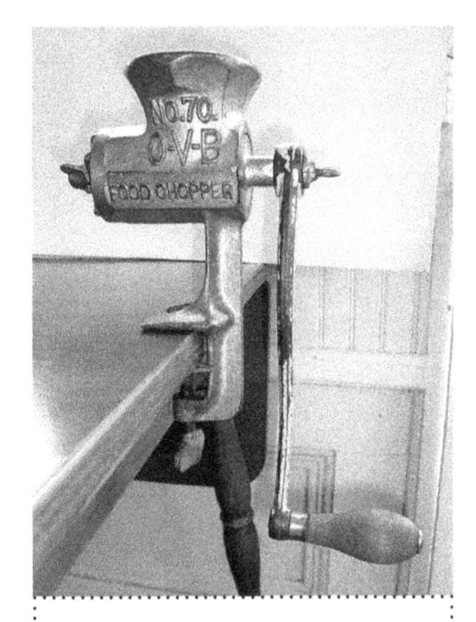


Meat grinders reduce large pieces into a soft, pliable mass into which spices, fats, or other additives can more easily be mixed. Grinders can be hand- or electric-driven and come in many different designs and shapes.

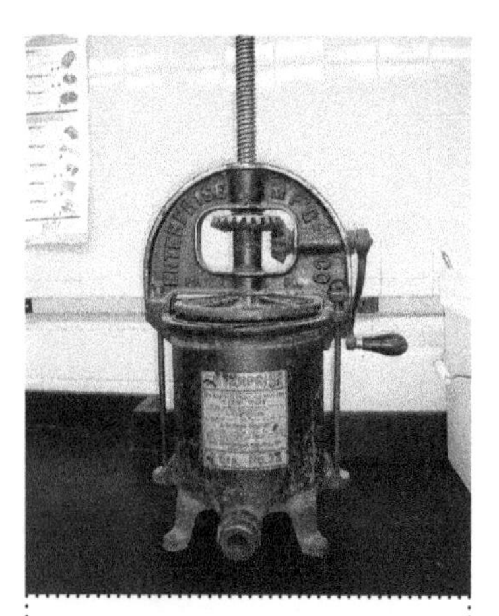

An old-style sausage stuffer was made of cast iron and operated by using a hand crank that pressed the plate, forcing the meat through the funnel opening at the bottom and into the casing.

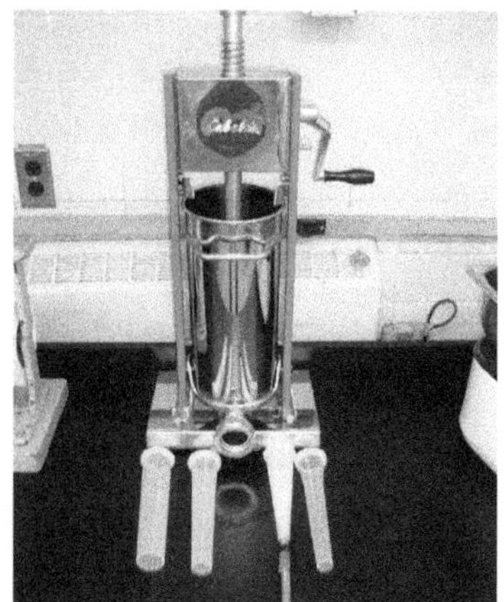

A modern sausage stuffer relies on the same principle as the old style, but it is made of aluminum or stainless steel. It is lighter in weight than its predecessor and has multiple funnels for different-sized casings. Be sure to thoroughly wash, sanitize, and rinse the stuffer and funnels or horns prior to and following each use and between different sausage-making sessions.

## SAUSAGE STUFFERS

You should consider buying a sausage stuffer if you plan on making your own sausages. There are several types available, including hand, push, crank, and hydraulic-operated by air or water. They can be made of plastic, stainless steel, or cast iron. Many small meat grinders are capable of supporting a small stuffing horn.

The piston-type stuffer is one of the most common for home use. It is operated by air or water pressure and will press the sausage quickly into the casing, producing fewer air pockets than hand-operated, screw-style stuffers.

1. Begin by placing the casing over the stuffer funnel or horn and tying the end. Use whatever function your stuffer has, hand crank or electric, and push the sausage into the casing.

2. Push until the desired amount is added; stop and give the casing a twist to separate it into links or small sections. Continue until the end of the casing and tie the end.

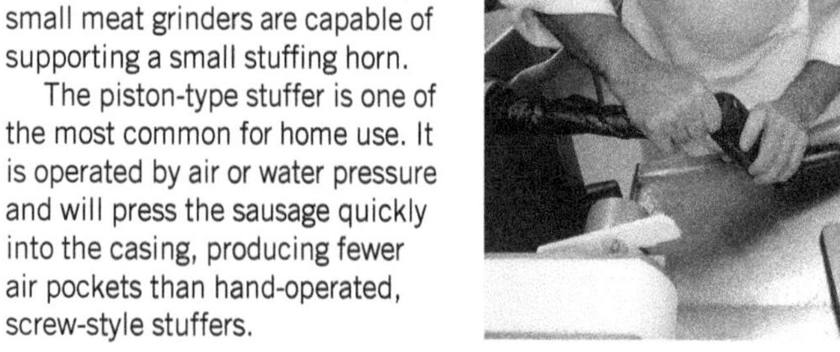

3. For larger casings such as summer sausage, use the same procedure by twisting each section after the desired amount has been added.

A push stuffer is quick to reload but has a small capacity. With this type of stuffer, you manually push down on a handle to force the meat into the casing. A crank stuffer has more capacity than a push stuffer and takes less effort to press the sausage into the casing because of the pressure created by your combined arm and screw action.

### SAUSAGE FUNNELS OR HORNS

The sausage funnel or horn constricts the movement of the sausage from the meat tub into the casing. As the casing fills, it pushes itself away from the funnel as it elongates. The size of the funnel is directly related to the size of the casing. Funnels are straight tubes, not tapered, and may range from 4 to 6 inches (10 to 15 cm) in length. To decrease the possibility of tearing the casing, coat the funnel with water or grease to help slip the casing over it.

## OTHER EQUIPMENT

There may be other items you'd like to keep on hand during your sausage processing, including measuring instruments such as a scale, measuring cups, and thermometers.

### SCALES

For weighing meats and other ingredients, a reliable scale is essential. A scale that measures both in pounds and ounces (kilograms and grams) should be sufficient for most of your needs. For recipes or curing chemicals where weights are measured in grams or ounces, a smaller scale may be necessary. If curing ingredients are being used, particularly sodium nitrite, it is very important to use a scale that can measure to the nearest tenth of a gram.

### MEASURING CUPS AND SPOONS

Measuring cups and spoons, ranging from 1/4 teaspoon up to 1 tablespoon (15 ml) and 1/4 cup (60 ml) to 1 cup (235 ml) for liquids or dry measure, will be useful for adhering to specific recipes. One simple rule of thumb is that smaller amounts are always easier to measure with spoons while larger amounts should be measured using cups for greater accuracy. Always measure the exact amounts called for, especially when processing sausages or other meat products.

### THERMOMETERS

Accurate thermometers are essential to help monitor and maintain appropriate temperatures during the processing and cooking of sausages. Handle all thermometers carefully and store safely. See pages 149 to 151 for a complete discussion of thermometers and instructions on calibration.

### INSTANT-READ THERMOMETERS

An instant-read thermometer is a probe containing two different metal coils bonded together. The coil is connected to the temperature indicator that expands when heated, moving the dial. Insert the probe about 2 inches (5 cm) into the center of the meat to insure a safe, accurate reading. Instant-read thermometers are used to assess when a specific temperature has been reached to assure safe eating and in making smoked sausage. They are good for use in sausage making because they can measure the temperature of a food within fifteen to twenty seconds. Although they are not used during the cooking process, they can be used at or near the end of it to check the final temperature. This will allow you to monitor the cooking progress without overcooking the product.

An oven thermometer can be set on one of the racks to monitor the temperature within the oven during the cooking time. One disadvantage may be if you do not have a window to check the thermometer, you may have to open the door to check the temperature, allowing heat to escape and prolonging the estimated cooking time.

Always sanitize thermometers before each use and when moving from one meat to another to avoid cross-contamination. The thermometers can be washed in hot, soapy water, then rinsed and dried.

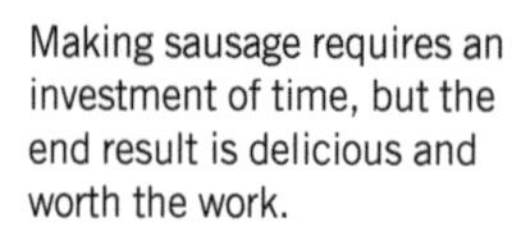

Making sausage requires an investment of time, but the end result is delicious and worth the work.

Once all the sausages are finished, they can be placed on racks for smoking. Regardless of the number of sausages, their size, or the type of smoker used, keep the individual pieces spaced so that they do not touch one another or any part of the smoking unit.

## SANITATION AND THE THREE Cs

Strict sanitation is critically important in sausage making and must be maintained to prevent bacterial contamination and food-borne illnesses. It is essential to handle raw meat in a safe manner that reduces the risk of bacterial growth. No meat product is completely sterile, but using proper procedures will minimize your risks. The most basic sanitation procedure involves using and maintaining clean surfaces before and after processing sausages.

It is easy to remember the three Cs of sanitation: keep it clean, cold, and covered.

### KEEP IT CLEAN

Wash all surfaces that you use with a diluted chlorine bleach solution of 10 parts water and one part bleach, as well as antibacterial soap. Nothing will replace vigorous scrubbing of the surface area with these products. This removes any grease or unwanted contaminants from the preparation area. Keep the area free of materials that do not relate to meat preparation or that will be used later but may accidentally come in contact with the meat. Utensils to be used and your hands should be thoroughly washed before beginning.

### KEEP IT COLD

Bacteria grow best in temperatures between 40°F and 140°F (4°C to 60°C). If you are cooking or cooling meat for cooked sausages, be sure your product passes through this range quickly because meat can be kept safe when it is cold or hot, but not in between. The meat you process should pass through this temperature range, whether being cooked or cooled, within four hours, but preferably less. This includes any butchering time involved. Cooling the fresh carcass is essential to a good meat product and is discussed elsewhere in this book.

During processing, cooked

### FOOD SAFETY TIPS

Always wash your hands.

Always wash your hands with soap and water before handling meat or beginning work. Rewash between tasks and after sneezing, using toilet facilities, or handling materials not part of your processing work.

Before and after use, thoroughly clean all equipment, knives, utensils, thermometers, bowls, and anything else used to cut or store meat. Clean and sanitize all surfaces that will be used.

Keep raw meat separate from other foods. Avoid cross-contamination between pieces of raw and processed meats. Avoid mixing of fluids and juices from other cuts or vegetables to be used.

Keep meat below 40°F (4°C) during processing.

Monitor temperatures at all stages of your processing.

sausages should have an internal temperature of 160°F (71°C), as this effectively kills pathogenic bacteria. Poultry meat should be cooked to 180°F (82°C) because of a more alkaline final pH. Ground meats are more likely to become contaminated than whole pieces because they have increased surface area exposure and go through more processing steps.

After cooking, you will need to cool the sausages quickly. This will prevent bacteria from attaching themselves and having an opportunity to grow while handling them.

The shelf life of any sausage has a limit. To minimize bacterial growth, you should store your sausage in a refrigerated or frozen state. Fresh or uncooked sausage can be kept safely refrigerated for several days while cooked sausages should be used within one week, unless frozen. Always remember that sausages are highly perishable products that don't get better with age. Never eat or serve sausage that has developed a slimy texture or an off smell. There is a good biological reason that your nose is placed near your mouth. If it doesn't smell right, it is better to discard it than risk eating it.

### KEEP IT COVERED

Meats, carcasses, and wholesale or retail cuts should be covered during any time you are not working on them. Your processing equipment should be properly stored in between use, as well as in any area used where butchering is done. Maintain screens, barriers, or traps to keep out vermin and reduce access for flies and insects.

## GAME MEAT SAUSAGE

While the majority of sausages made use pork or beef separately or in combination, game meat can be used and substituted instead to create unique and original flavors. Most any meat from wild animals can be used, but, like meat from domesticated animals, it is important that it is handled properly after the animal is killed. The same awareness of temperatures is required, and dressing the animal as soon as possible while keeping the meat under 40°F (4°C) will help limit bacteria growth, reducing the chance of a food-borne illness.

Game meat is typically aged after it is dressed to increase tenderness. This is not a necessary step if the meat is to be used for sausage because it will be tenderized and broken down through the grinding process. Using the less tender cuts and trim pieces from wild game in sausage increases the volume available.

Game meat contains a distinct flavor that comes from the fat and not necessarily the meat itself. Removing all external fat prior to grinding will allow you to process game meat in the same way as beef or pork. However, game meat is leaner and contains less fat in the muscle. This will make a dry and unpalatable sausage unless you add unsalted pork fat when grinding. The pork shoulder butt is often used as a fat for game sausages. Generally, you will need to mix in a fat content of 15 to 20 percent to have a desirable flavor and texture. Blending different meat and fat percentages will affect your final product. Experimentation will be the best way to discover what you like best.

CHAPTER 11

# MAKING JERKY

**AS WITH ANY NEW HOBBY, IT MAY BE TEMPTING TO JUMP RIGHT INTO MAKING JERKY, LEARNING AS YOU GO AND NOT FULLY UNDERSTANDING THE BASIC STEPS. HOWEVER, THERE'S A REASON JERKY MAKING IS OFTEN NOT PART OF MORE GENERAL PRESERVATION BOOKS. THERE ARE ADDITIONAL CONSIDERATIONS AND VERY CLEAR FOOD SAFETY RECOMMENDATIONS THAT MUST BE FOLLOWED TO ENSURE A SAFE FOOD PRODUCT.**

I'm sure you agree, it is worth the time it takes to think about and learn the basic principles of food safety. After all, nothing is more important than you and your family's health. These basics include sanitation; factors affecting moisture, oxygen, temperature, and time; preventing and retarding bacterial growth; and muscle basics that can contribute to jerky's overall quality.

In addition to understanding good food-processing techniques, you should develop a plan for acquiring the tools and equipment needed to adequately process the meat into a safe, quality finished product. Your plan and equipment needs should be thought out long before you have the meat in hand.

The range of meats you can use for jerky is large. In fact, it's limited only by your imagination, interest, and willingness to experiment. Hunters typically have a wider range of animals to choose from because they are out in the wild. For those who don't hunt, the range of meats available is still extensive. Some wild and domestic meats that can be used include elk, moose, deer, cow, bison, antelope, rabbit, squirrel, sheep, goat, swine, and reindeer. Poultry and waterfowl, whether wild or domestic, can include chickens, turkeys, geese, ducks, and the ratites, ostrich and emu.

The most common variety of jerky—beef jerky—is easily sourced from a reputable butcher. Other meats at the butcher or a local quality supermarket can also be considered.

Making jerky is relatively simple and can be done by anyone who has access to a kitchen and a dehydrator or smoker. You don't need any special expertise, but you do need to understand and follow some easy directions. Like homebrewing or canning, making jerky at home requires attention to a few basic principles to ensure a safe, stable food product.

Making your own jerky allows you to choose from a wide variety of meats, such as beef, chicken, or wild game. It also puts you in control of the kind and intensity

**NOTE:**
One other significant consideration to be made once the meat is ready to be processed is to cook it or heat it to the appropriate temperature. For temperature information, see page 201.

Cleanliness is very important when handling raw meat, regardless of if it is sourced from a store or the wild. Safe food processing begins with washing your hands thoroughly with soap and water prior to handling any meat.

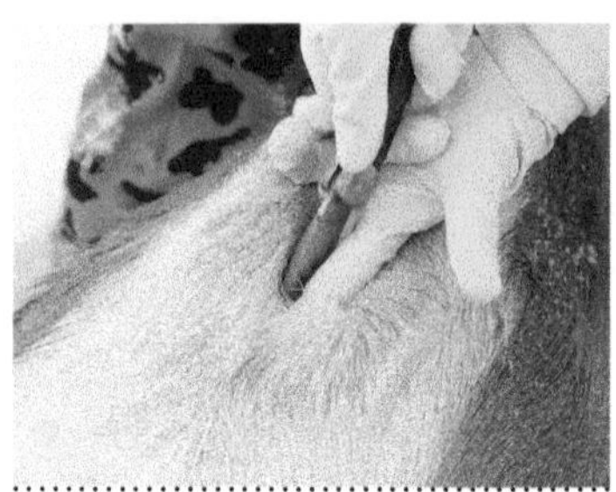

Use rubber or latex gloves when handling or field dressing wild game. Gloves act as a barrier between your skin and an animal's body fluids in case of any unseen infections.

of flavors in your jerky, and it allows you to create a high-quality product without chemical stabilizers or preservatives. This chapter will walk you through the steps required to produce safe, top-quality jerky and will discuss some of the equipment you'll use.

# JERKY AND FOOD SAFETY

The most important aspect of making jerky is to consider all the points where meat contamination may occur and to ensure contamination doesn't happen. Preventing food-borne illnesses should always be your top priority.

Unless you're making vegan jerky, you will start with raw meat or fish either harvested in the wild or purchased at a local meat counter. There are basic steps you can take to ensure safe handling of raw meat cuts that will minimize the risks involved with processing it. You may already be familiar with the three standard "Cs": keeping meat **C**lean, **C**old, and **C**overed. There's one additional "C" I like to include as well: keeping meat separate to prevent **C**ross-contamination.

## REMEMBER THE THREE Cs

The Three Cs cannot be overstated and they are as important when making jerky as with other meat products. The more a piece of meat is cut up and deconstructed, the greater the risk of contamination because of the increase in surface exposure of the pieces.

Again, let's start with keeping things clean. Strict sanitation is required before and after cutting any meat—whether in the wild or in your home—and processing it into jerky to prevent bacterial contamination and food-borne illnesses. It is especially important to handle raw meat in a sanitary environment to reduce the risk of bacterial growth while it is at room temperature. No meat is completely sterile, but using proper procedures will minimize your risks.

The greatest challenge for maintaining sanitation occurs when harvesting and field dressing game animals in the wild. Domestic animals are raised under controlled environments that have protocols to guide producers in raising healthy animals for the market. Their meat is processed under strict sanitary conditions before it is sold to another market or grocery store. Wild game does not have such guidelines and health regulations.

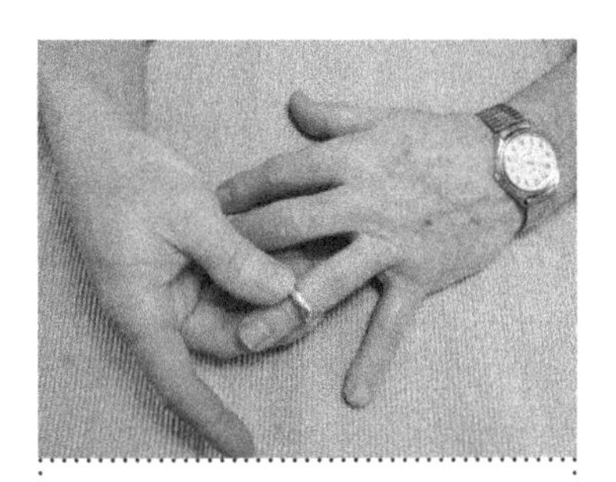

Remove all rings and watches from your hands and wrists before working with raw meats to avoid the introduction of bacteria.

Cleanliness and sanitation also play a significant role when indoors. This includes using and maintaining clean knives and work counters or cutting surfaces; ensuring proper insect, fly, and rodent control; and frequent hand washing.

Cleaning is the first step before sanitizing. Cleaning involves removing any organic matter by using detergents, solvent cleaners, acid, and/or abrasive cleaners as necessary. Sanitizing follows and is the application of heat or chemicals, typically a chlorine-based solution, to surfaces that will come in contact with food.

Whatever you use for a cutting surface must be cleaned thoroughly and then sanitized before placing any meat on it to prevent contamination. Vigorous

## FIELD DRESSING AS CLEANLY AS POSSIBLE

For many years, it was thought that the muscle of an animal was sterile if it had not been injured, cut into, or bruised. However, recent research has found viable bacteria within muscle tissue. This means that when you harvest an animal, especially in the wild, extreme care must be taken to prevent the introduction of foreign bodies into the carcass through your actions. This care begins with the knives and equipment you use during the meat processing and continues until the meat has been dried for use.

Successful hunting typically results in wounds made from bullets, arrows, or hooks. These create openings in muscles and body cavities, which can become contaminated. Portions of the carcass may have materials embedded such as hair, dirt, metal shards, and any mixture of blood, bone chips, and fecal matter. All damaged tissue needs to be cut out and removed. Your knife should be sterilized before and after excising the wound area or digging out a bullet or arrow tip. Be aware that bullet or arrow fragments may splinter and be embedded in areas not readily observed and may pass deep into muscles without leaving a noticeable path. These pieces may be too small to be seen and may be dispersed away from the wound channel. Any wound area should be excised as soon as possible to prevent bacterial growth from spreading into the muscles.

Still, the most exposure a wild game carcass has to its environmental conditions occurs during field dressing. Whether it is a deer, squirrel, elk, moose, or rabbit that has been killed, its surroundings typically include dirt, dust, plants, insects, and depending on the time of year, atmospheric temperature concerns. It is vitally important to minimize carcass contamination from the elements present by using as many precautions as possible. This could include using plastic sheeting on which to lay a carcass while making any initial cuts. Try to refrain from using water from creeks or streams to flush out any body cavity. Water in the field may contain bacteria or soil particles that can contaminate the meat and lead to an excellent environment for bacterial growth.

If you decide to or need to field dress your game animal, you should always wear protective gloves, either rubber or latex, especially when dressing and skinning some game such as wild rabbits or hares. This will protect you from possible contact with a communicable disease known as tularemia, or rabbit fever. This is a very infectious disease that can be transmitted from one rabbit to another by lice or ticks, or to humans by handling the flesh of an infected animal, inhaling the bacteria during the skinning process, or through a tick bite. Since its entry is through cuts, abrasions, or inhalation, you need to take precautions. You may not recognize an infected animal prior to skinning it, so strict sanitation is essential.

Keep your work area clean and free of any items not involved with your food processing. A simple cleaning solution can be used.

Wash all knives that you will use with hot, soapy water or other cleaning solution and then rinse with clean, clear water. Be sure to thoroughly clean the joint where the blade attaches to the handle.

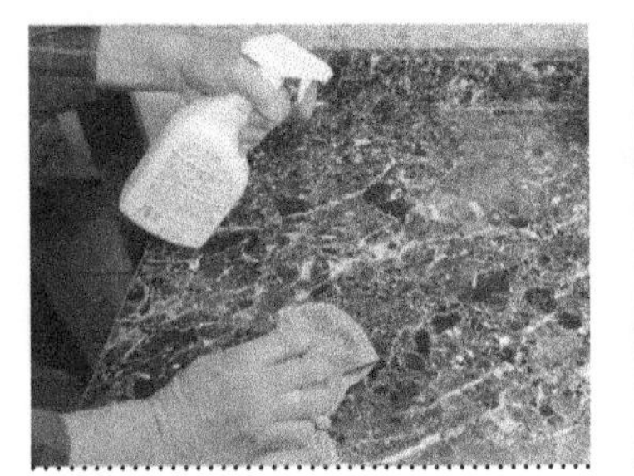

Clean all work surfaces thoroughly and wipe dry before placing raw meat on them.

**NOTE:**
It's always a good idea to remove any materials or equipment that will not be used in the meat preparation from your work area. If they are not being used, they may become a hindrance in your work space and should be treated as contaminated if you have to move them while you are cutting the meat.

scrubbing may be necessary and cleaning products should remove any grease or unwanted contaminants from the preparation area before sanitation. If they don't, you should repeat the cleaning process. All the utensils that you will use should be cleaned before they're sanitized as well.

For a sanitizer, mix a solution of 1 part chlorine bleach to 10 parts water and use it to wash the surface of your table, cutting board, or counter. You can use this solution to briefly soak any tools, knives, or other equipment that will contact meat and then rinse these with clear, clean water before use.

Another option is to use an acid-based, no-rinse sanitizer such as Star San. This is a commercially available cleaner that is odorless and flavorless and will eliminate any concerns about tainting your meat. Although chlorine reacts quickly and becomes inactive quickly, you should always follow the mixing directions exactly. Spray the surface of your table, cutting board, or counter with this solution and use a small bucket or bowl to soak any tools.

Any knife being considered for use in cutting meat should be thoroughly washed before and after each use. Use mild soapy water and clean by hand. When cleaning knives that are more than one solid piece, you should pay close attention to the area where the blade attaches to the handle. This is an area where meat or blood residue will remain after cutting and is an ideal habitat for microorganisms to grow later as knives are generally stored at room temperature. Remember, if it's not clean, it can't be sanitized.

Other equipment that may be used, including slicers, meat grinders, extruders, rolling pins, presses, scales, or other utensils, should also be thoroughly washed prior to each use. They can also be sanitized in a chlorine bleach solution as an added precaution, but be sure to thoroughly rinse each piece separately in clear, clean water before using to remove any possible chlorine tainting of the meat.

The type of material you use to clean counter or cutting board surfaces can be a concern. Sponges and wet cloths are typically used for cleaning and when used properly can do an excellent job. However, bacteria live and grow in damp conditions, and wet sponges and cloths make ideal harbors for foodborne pathogens. Sponges and cloths that give off unpleasant odors are a sure sign that unsafe bacteria are present.

Sponges can be used successfully if properly cleaned. You can't eliminate all of the bacteria that may reside in sponges, but you can reduce the risk of cross-contamination and the spread of harmful bacteria by following a few easy steps.

- First, clean your sponges daily. It will lower the risk of bacterial growth. The USDA has found that by microwave-heating damp sponges for 1 minute or dishwashing with a drying cycle will kill over 99 percent of the bacteria, yeasts, and molds present. You can also disinfect sponges with a solution of 1/4 to 1/2 teaspoon of concentrated bleach per quart (950 ml) of warm water and a soak of a minimum of 1 minute.

- Second, replace sponges frequently. This will reduce bacterial growth. Frequent use of any sponge can develop pockets that harbor bacteria even when routinely cleaned. Don't wait until the sponge falls apart before tossing it away. A clean sponge is a safe sponge.

- Third, store your cleaning sponges in a dry place. Wring them out completely after each use and wash off any loose food or debris. Don't allow them to remain wet on a countertop because this will mean they will dry more slowly, which will allow bacteria to multiply quickly. Don't store damp sponges under a sink or in a bucket where they won't be exposed to circulating air.

Dishtowels and cloths should be handled in the same way as sponges. Keeping them disinfected and dry, while periodically replacing them, will help keep bacterial growth to a minimum. You should also use separate towels or cloths: one for drying hands and one for wiping counters. Remember to wash them in hot water and dry them on high heat in your dryer.

Whenever possible, you should clean up spills with disposable paper towels rather than using sponges or cloths for such small tasks to avoid cross-contamination.

Once everything is ready, it's a good idea to thoroughly wash your hands with soap and water before touching any meat or clean work surfaces. Remove any rings, jewelry, or other metal objects from your hands, ears, or other exposed body parts before cutting meat. Always rewash your hands between tasks, as well as if you come into contact with anything unsanitary: if you sneeze, use the bathroom, or handle materials not part of meat processing, you should rewash your hands.

## KEEP IT COLD (AND/OR HOT)

Mismanagement of temperature is one of the most common reasons for outbreaks of food-borne diseases. Meat can be kept safe when it is hot or when it is cold, but not in between.

Store your raw meat in a refrigerator until you begin processing it.

Prior to processing or dehydrating your jerky, it is likely you will be working with a raw meat product for a period of time. The less time you subject the meat to room or ambient temperatures, the less risk there will be in it harboring harmful microorganisms that cause spoilage. This is especially true if the meat in question has been harvested in the wild where temperature and time will affect its quality more than if purchased in a store.

If meat is stored below 40°F (4°C), most of it can be kept safe from harmful bacteria for a short time. When frozen, most microorganisms that are present are merely dormant and can revive when thawed. If you have thawed the meat you plan to use for jerky, it all should be processed as soon as possible and not refrozen to use for jerky later.

Temperature remains the critical factor once you proceed to jerky production. In decades past, the standard was for the meat to be heated between 130 to 140°F (60°C), which many dehydrators could achieve. However, the current United States Department of Agriculture (USDA) recommendation for safe jerky making is to heat meat to 160°F (71°C) (and poultry to 165°F [74°C]) before the dehydrating process begins. Reaching these temperatures will assure that the wet heat will destroy any bacteria present. Research has shown that without reaching these temperatures before dehydrating, any bacteria present after drying become more heat resistant and may survive in sufficient quantities to create health problems later. We'll learn more about this on page 190.

## KEEP IT COVERED

All foods have a diminishing shelf life after they are opened or made, even if properly stored. However, the manner in which you cover, contain, or wrap foods prior to use or storage determines how well it keeps in a refrigerator, cupboard, or freezer.

Temperature has the greatest effect on meat, especially as it increases. As previously mentioned, raw or cooked meat should be kept chilled until used. Even your refrigerator will not have consistent temperatures throughout. Interior drawers tend to have slightly higher temperatures than shelves. The door shelves are also generally warmer because of their exposure to room temperatures once they are opened. So it's best not to store highly perishable foods such as meat in the drawers or doors of your refrigerator.

Wrapping or enclosing foods in containers will help keep them fresher than if left uncovered in the refrigerator. Uncovered meat has more exposure to oxygen, which can cause bacteria to multiply faster. Oxygen also tends to dry out the meat surface much quicker.

Wrapping or containing meat serves several purposes. It forms a barrier between the meat and oxygen, it prevents refrigerator odors from transferring from one food to another, and it helps prevent cross-contamination between foods that may occur through drips or touch. Outside the refrigerator, wrapped or covered meats are less exposed to flies, insects, or pets.

There are a number of products you can use to wrap, cover, and store your foods, and each has its advantages.

- Aluminum foil is good for keeping moisture out of food, and it protects food from light and oxygen. It has reactive properties, however, and shouldn't be used with acidic foods such as berries or tomatoes. A layer of plastic wrap followed by a layer of aluminum foil will provide a double protective layer for any food. It is also very good for use in freezing foods.
- Plastic wrap provides close to an airtight seal on bowls and containers without lids. It has a unique ability to adhere to a variety of surfaces, whether plastic or glass. Because it is transparent, you can see what's inside without having to open the package. Don't use or leave plastic wrap on any container or food being heated or cooked.
- Resealable plastic bags come in a variety of sizes and weights (thicknesses). The heavier-weight bags are good for freezing foods. They provide the best protection when the air is pressed out of the bag prior to sealing it.
- Airtight glass or plastic containers with lids can be cooled or frozen. Some, such as Pyrex containers or other shatterproof types, can be used in either cold or boiling water.
- Freezer paper is a plastic-coated paper designed for wrapping foods destined for the freezer. It is much heavier material than aluminum

foil or plastic wrap and provides good protection for storing foods. You can also write on it so that you know what is inside.

- Waxed paper is a moistureproof material that is made by applying wax to the paper surface. It has a slightly higher heat tolerance than plastic wrap but should not be used in cooking or baking. Its nonstick surface properties make it a good barrier between frozen food cuts wrapped in the same package.

- Parchment paper is often used in heating or baking because of its ability to withstand oven heat without combusting (except if it comes into contact with the burner coils). It can be used to act as a barrier between foods you are working with on your table or cutting board, or between cuts of meats you want to freeze in the same package so they will separate with little trouble. Parchment paper can be used to line pans or trays for quick clean up.

- Vacuum sealers are appliances that remove all the air from any package being sealed, which will extend the shelf life of the food inside (not indefinitely) and decrease the chance of freezer burn. Vacuum-sealed packages can be stored in your refrigerator, freezer, or cupboard, depending on the type of food sealed.

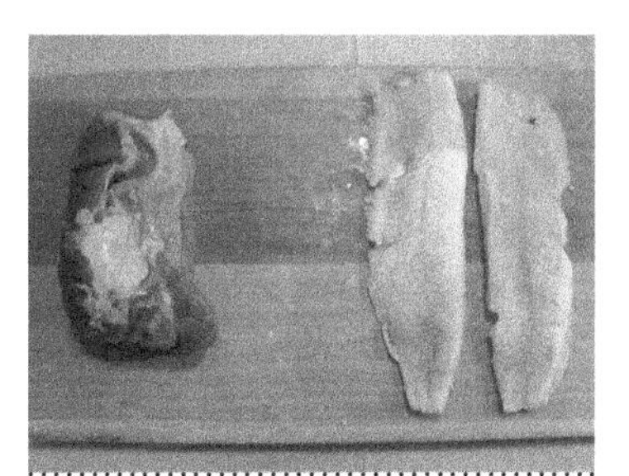

Prevent cross-contamination by keeping different foods, such as meat and fish, separate from each other. Always clean the surface worked on prior to placing a different food on it.

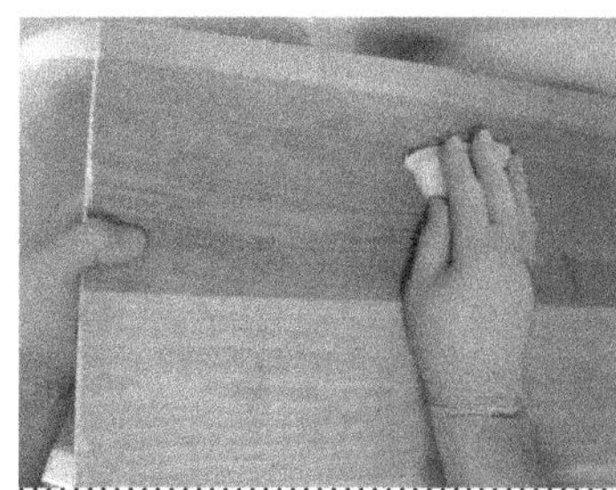

If using a wood cutting board, make certain that it is free from cracks or crevices that may harbor bacteria. Thoroughly clean your cutting board before and after each use.

## PREVENTING CROSS-CONTAMINATION

Cross-contamination occurs when one food comes in contact with another, creating the potential of spreading bacteria from one source to another. One or both foods might be raw, or one can be raw and one cooked. Cross-contamination can occur when foods touch, or it can occur when surfaces have had mutual contact with several foods, such as when one food is placed on a plate, counter, or cutting board, then removed, and then another food is placed on the same now-contaminated surface. It can also occur between knives that have not been cleaned between uses, and even in your grocery cart if juices happen to leak from one package to another.

There are several steps you can take to avoid cross-contamination of any foods.

- Always wash your hands thoroughly with warm, soapy water prior to, during, and after handling raw meats and other foods. Make sure all counters, cutting boards, plates, knives, and other utensils are thoroughly washed and dried with clean towels.

- Separate different foods into different dishes, plates, or bowls prior to use.

- Keep raw meat, poultry, seafood, or eggs on the bottom shelf of your refrigerator and in sealed containers or bags so they cannot leak or drip onto another food.

- Use a clean cutting board for each of the different foods you are working with. Use separate boards for raw meats, vegetables, and other foods. If you are using the same knives or equipment for all your cutting and processing, wash them thoroughly each time you move from one food to another or from one cutting surface to another. Replace any cutting boards that have cracks, holes, or grooves, as these are good places for bacteria to hide and grow.

- Avoid using leftover marinade for any other meats. If you need the same marinade for another dish, set aside a small amount before placing raw meat in it. This will leave you with a fresh marinade for later use.

- Clean your refrigerator shelves on a regular basis, particularly if juices from raw meat, vegetables, or seafood have leaked, dripped, or spilled.

- Try to avoid mixing raw meat, vegetables, seafood, and eggs in the same bags when you check out at the grocery store. Try to separate frozen and fresh food into separate bags.

- Never place cooked food on a plate that was used for raw meat, poultry, seafood, or eggs.

## PREVENTING AND RETARDING BACTERIAL GROWTH

Although sanitation and the three "C" dynamics have become routine components of jerky processing, sausage making, and carcass deconstruction, misuse or incomplete application of any one of them can be detrimental. Preventing and retarding the development of harmful organisms should be your primary objective while handling raw meat and turning it into jerky. Consuming microorganisms that have grown and propagated in meat can cause serious illness or even death. This concern should not be taken lightly. When health problems surface relating to eating meat products, even if it's jerky that has been dehydrated, they are generally a result of intoxication or infection.

Intoxication occurs when heating or processing fails to kill the microbes in food. Those that are able to survive can produce a toxin that, when eaten by humans, can produce illness. In undercooked meat, for example, infection occurs when an organism such as salmonella or listeria is consumed.

There are several types of toxins, including exotoxins and endotoxins.

- Exotoxins are located outside of the bacterial cell and are composed of proteins that can be destroyed by heat through cooking. Exotoxins are among the most poisonous substances known to humans. These include *Clostridium botulinum*, which causes tetanus and botulism poisoning.

- Endotoxins attach to the outer membranes of cells but are not released unless the cell is disrupted. These are complex fat and carbohydrate molecules, such as *Staphylococcus aureus*, which are not destroyed by heat.

Not all bacteria are bad, however. According to one New York University study, the human body may carry as many as 180 different kinds of bacteria on its surface. Although molds and yeasts can affect meat quality and cause spoilage, their effect is far less significant or life-threatening than toxins or bacteria. Molds typically cause spoilage in grains, cereals, flour, and nuts that have low moisture content or in fruits that have a low pH. Yeasts will not have a significant effect on meat because of the low sugar or carbohydrate content of muscle. They need high sugar and carbohydrate levels to affect a change.

Several parasites may cause problems if the meat being used for jerky is undercooked or improperly processed. A parasite infection occurs first in the live animal and then, after butchering, may be transferred to humans while still in an active state. There are three parasites that you should be aware of, depending on which meat you decide to use for jerky. These include *Trichinella spiralis*, *Toxoplasma gondii*, and *Anisakis marina*. Learn more about these on page 18.

One disease that may be a concern is chronic wasting disease (CWD). It is a progressive, fatal illness in deer and some other herbivores. It has attracted attention because it has been identified in animals in fifteen United States and two Canadian provinces. CWD is believed to be caused by a prion protein that damages portions of the brain in affected animals. It causes progressive loss of body condition, behavioral changes, excessive salivation, and finally, death. The mode of transmission is not fully understood, but it is thought that the disease is spread through direct contact between animals or exposure to contaminated water and food supplies.

As of this writing, no strong evidence of CWD transmission to humans has been reported. Still, hunters harvesting animals originating in known CWD-positive areas should have them tested before consuming any of the meat, regardless of whether it will be made into jerky. You can take precautions prior to harvesting an animal by not shooting, handling, or eating any deer that appears sick or decimated or tests positive for CWD. Also, if you field dress one of these cervids, it is a good precaution to wear gloves, bone-out the meat from the carcass, and minimize handling of the brain and spinal cord tissues (for more on this, see clean field-dressing practices on page 185).

## MOISTURE AND OXYGEN

Moisture in meat is essential for palatability but is also a medium for microbial growth. The level of moisture in fresh meat is high enough to provide spoilage organisms with an ideal environment for growth. Researchers have found that moisture levels of at least 18 percent will allow molds to grow in meat. Drying meats through a smoking process or by making it into jerky will typically eliminate any concerns with moisture.

Oxygen is an unwelcome agent when processing meats. Yeasts and molds are aerobic microbes that need oxygen to grow.

Drying is the safest procedure to follow when making homemade jerky because it acts as an inhibitor of enzyme action by removing moisture. When moisture is removed, enzymes cannot efficiently contact or react with the meat fibers or particles. Without this interaction, bacteria, fungal spores, or naturally occurring enzymes from the raw meat cannot grow to proportions that can cause severe illness. Minute traces may still be present, but with no growth, they lie dormant.

However, lying dormant does not mean they can't resume growth if favorable moisture or temperature conditions are introduced. This may occur if jerky is left out in moist conditions and is one important reason to keep any homemade jerky in cool, dry conditions until eaten.

Eating jerky that has been made with sanitary practices carries minimum risks, for several reasons. First, the high internal temperature that is created (to be discussed in detail later) significantly reduces the survival of salmonella, E. coli, trichinosis, and other bacteria. Additionally, processing jerky typically involves using more salt than you would use with many other uncooked foods for your table. Salt acts as an inhibitor of bacterial growth but also adds flavor after the meat is dried. Third, drying can eliminate more than 90 percent of the meat's water content, a medium needed for bacterial growth.

## MUSCLES AND MOLECULAR TRANSFORMATION

The position of the muscles on the skeleton has a significant impact on the texture of meat. For example, muscles that create movement—such as the front and hind quarters in herbivores such as cattle, pigs, and sheep, or the wings and thighs of wild turkeys—receive more exercise than the loin or breast areas of these animals and fowl. The more exercise or movement a muscle uses, the more blood flow is needed. This, in turn, creates a darker color of meat because of the flow of hemoglobin needed to deliver oxygen to the muscle. The more hemoglobin (sometimes referred to as myoglobin) a muscle contains, the darker color the muscle will be. In fish, the large muscles of the body and tail comprise the majority of the body mass and because they provide the most movement, they will contain the most blood. The muscles that receive the most use for movement typically also contain the least amount of fat because the fat is synthesized by the muscle for energy. Muscles with little fat also tend to be less tender in texture and, ultimately, in taste. Although any meat cut can be used, jerky is often made from the less desirable cuts as a way of using rather than discarding them.

The muscles of a harvested animal or bird, whether field dressed or butchered in a confined indoor area, go through a molecular transformation once the heart stops. This can influence the muscle texture. With the cessation of blood and oxygen flow, the muscle pH begins to gradually drop. This occurs because the glycogen reserves within the animal's muscles are depleted and are converted to lactic acid. Lactic acid levels rise, the pH begins to drop, and the reserves of creatine phosphates diminish. Creatine phosphates aid in muscle movement; when they are no longer available, the muscle filaments can no longer slide over one another and the muscle becomes still and rigid, resulting in a condition known as *rigor mortis*.

Soon after an animal is harvested, the muscle's normal pH declines from 7.0 to 5.5 as a result of the loss of glycogen held in the muscle and its conversion to lactic acid. The degree of acidity or alkalinity (pH) will have an effect on the growth of microorganisms. Most of these will thrive at a pH that is nearly neutral (7.0), better than at any other level above or below it. Meat pH can range from 4.8 to 6.8, but microorganisms generally grow slower at a pH of 5.0 or below. This acidity level can act as a preservative in some instances and is generally not a concern unless there is a long delay in processing the carcass at room temperatures.

The amount of time it takes an animal's muscles to reach their final pH levels is influenced by several factors. These include the species, cooling rate of the carcass, and the extent of the animal's struggle at the time of death. Deer muscles take longer to reach their final pH level than many other wild animals. Cooling affects the time because metabolism is slowed when the carcass is subjected to lower temperatures. Finally, the animal's activity level immediately prior to the killing will affect the pH, with less activity prolonging the period of pH decline.

## TOOLS AND PREPARATION

You have learned to keep your raw meat safe from harvest to processing. You have learned to avoid contamination, spoilage, and bacterial growth that affects quality and health, and why meat decreases in quality as time goes by. What else do you need to know? Understanding your work and storage space needs and the equipment you will use will get you ready for processing the meat.

The good news is that elaborate working areas are not required to make jerky. A solid table with sturdy legs or a stable counter space can be made into suitable areas to cut meat. Whatever surface area you use, make sure it can be easily cleaned and that it's made of nonporous material that will not harbor food residue. Using a hard, even surface will make cutting meat easy and safe.

Home-use cutting boards are typically made of nonmetallic materials such as solid plastic, marble, glass, or rubber, and are generally corrosion-resistant. Their nonporous surfaces are easier to clean than wood, but if kept in good condition, wood boards can work as well.

The USDA Food Safety and Inspection Service recommends that all cutting boards used in your home first be cleaned with hot, soapy water before and after each use and then rinsed with clear water and either air dried or atted dry with clean paper towels. It also recommends that separate cutting boards be used for meats and vegetables.

Your preparation should also include gathering cleaning and sanitizing supplies. There is a wide variety of products available and there is no single detergent that is capable of removing all types of soils or complex films that may be a combination of food components, surface oils, or dust. Common cleaning agents include detergents such as dishwashing liquids that alter the physical and chemical characteristics of the substances being cleaned to neutralize and degrade them; solvent cleaners such as ammonia that contain grease-dissolving agents; acid cleaners such as hydrochloric acid that can remove mineral deposits that alkaline detergents cannot; and abrasive cleaners such as fine steel wool, copper, or nylon. Sanitizing compounds include antiseptics that are used against toxic agents or bacteria that may cause infections or have a putrefaction effect; disinfectants or germicides that are applied to stationary objects such as floors and countertops and will kill vegetative cells but not spores; and bactericides that can be used to kill certain groups of microorganisms but, depending on the type or strain, may only prevent their growth.

Sanitizers that include chlorine bleach, hydrogen peroxide, or white distilled vinegar, either in diluted or undiluted concentrations, are effective on food preparation surfaces. However, don't use baking soda because it is not an effective sanitizer at any temperature, time, or concentration. Using a combination of heat and chemicals is the best way to make food preparation surfaces and equipment safe and to help avoid food contamination.

For making jerky, there are a few essential tools you need to buy if you don't already have them in your kitchen. We'll cover the key items in detail on the pages that follow. You'll also want to make sure your kitchen has the following basics: measuring cups and spoons, glass bowls or other nonmetallic containers for marinades, containers or other packaging materials for finished jerky, plastic bags or plastic wrap, and kitchen towels and paper towels. Also, if you're planning to use your oven in the jerky-making process, you'll need to make sure you have clean and sturdy oven racks as well as cookie sheets and foil.

Of course, you'll also need to decide whether you'll be using your oven, a dehydrator, and/or a smoker. The tools you use for drying are tightly integrated with drying techniques. We will discuss them starting on page 211. No matter which method you use, though, you will need the tools and equipment on the following pages.

## COLD STORAGE SPACE

As already discussed, temperature is a primary factor you should control during meat processing and preservation. You should make sure the meat you use, whether harvested or purchased, is kept below 40°F (4°C) until your processing begins. For small batches of jerky, you should have no problems using your fridge. However, should you scale up or bring home a large amount of meat from a hunting trip, you need to make sure you have the dedicated fridge space ready.

Although properly processed jerky needs little refrigeration, it is advisable to keep your finished product in a cool, dry place. Keeping jerky in a refrigerated or frozen state will lengthen the time it will be available to you. To keep it for an extended period in a freezer, it will be best to have it vacuum packed to reduce the formation of ice crystals, which can reduce the meat quality. A vacuum packer removes the air from the package prior to sealing, which prevents the invasion of moisture, so it can be a good way to extend the life of jerky even if you don't plan to freeze it. Remember that even if you freeze jerky, improper jerky processing may not have removed all bacteria from the meat. The harmful bacteria may be reactivated once the jerky is thawed out. It is essential that only properly processed jerky meats be frozen for consumption later.

Nonmetallic containers, glass bowls, measuring cups and spoons, and paper towels are some items that will be useful while making jerky.

# CUTTING WHOLE MEAT JERKY

Before we go through specific meat sources, let's address the three primary ways you'll slice and process meat to break it down into smaller, more manageable pieces. The most common way to do this for jerky is slicing into uniform strips, so we'll cover that first. Then, we'll discuss the differences in technique used for making chunk and ground-meat jerkies. Keep in mind that you may develop techniques that work better for your particular situation or meat; these are only guidelines. Also, while the slicing information in particular was written particularly in regards to using beef, the most popular jerky meat, the basics can be used for many other animals including bison, chicken or other poultry, turkey, pig, sheep, lamb, and goat.

## BEFORE YOU SLICE . . .

The meat cuts you select for jerky should be as lean as possible, meaning you will need to trim off as much fat as you can manage before proceeding to processing. Although fat adds flavor and juiciness to meat during a regular cooking process, it can be detrimental to jerky quality. The fat in meat can turn rancid and produce an off flavor if the meat is not eaten in a short time. This is one reason the top and bottom rounds of the hind quarters are often used for jerky making: they contain less fat. It is important to remove as much fat as possible if grinding the meat because the fat will become dispersed within the ground meat mixture. The less fat that is interspersed in ground meat, the better the jerky will dry.

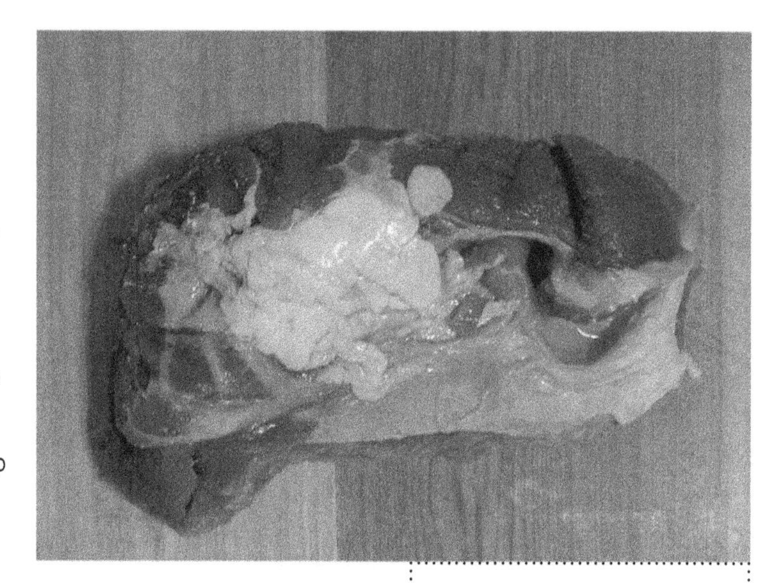

1. Buy as lean a cut of beef as possible for your jerky. Meat markets tend to leave some fat on the cut to add weight and enhance its appearance.

Prior to cutting or slicing the meat, you should keep it very cold in your refrigerator or even just below freezing (30°F [−1°C]). This will firm up the muscle and make it easier to slice, either by knife or with a slicer. If possible, use a mechanical slicer, as this will create more uniform pieces than using a knife. A steady hand with a sharp knife can suffice.

Muscle fibers lay in bundles in various configurations, which gives them a striated appearance. These fibers form the basic mechanism that controls muscle contraction and movement. Skeletal muscles are most commonly used for jerky making. They are covered with a dense connective tissue sheath called the *epimysium*. Each of these muscles is divided into sections, called bundles, by a thick connective tissue layer called the *perimysium*. Clusters of fat cells, small blood vessels, and nerve bundles are found in this layer. The fat cells will appear white, most often as streaks through the meat or surrounding the bundles (what most people call marbling).

2. Begin by slicing off as much surface fat as possible. Fat will increase drying time for the meat and can cause off-flavors in storage if too much is dried along with the meat.

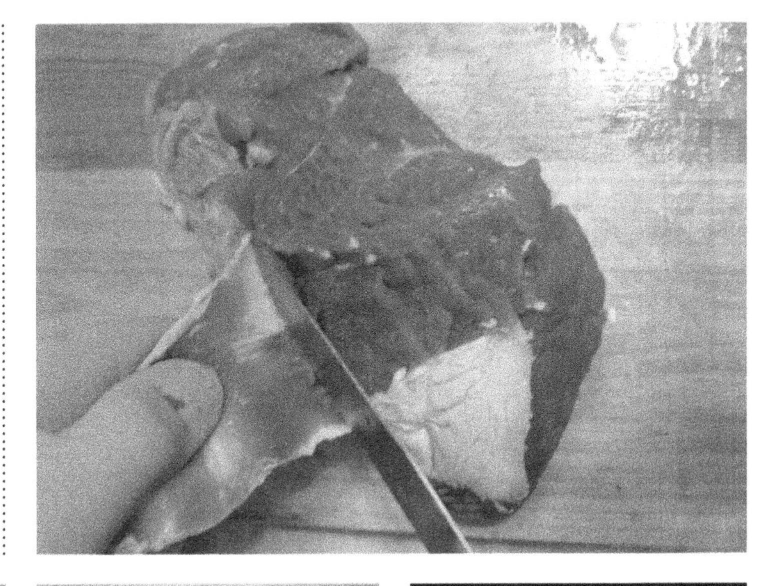

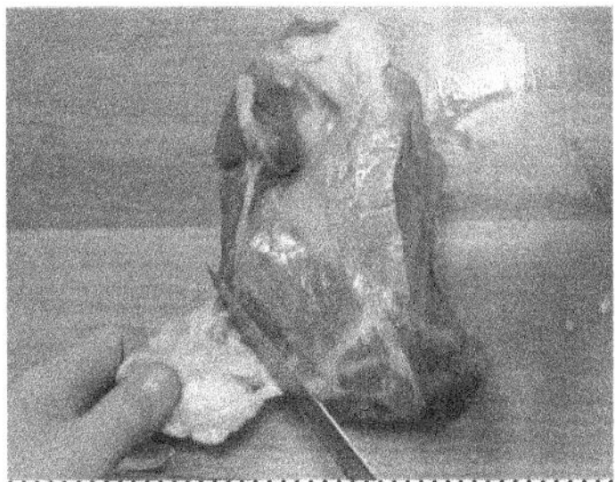

3. The silverskin that may cover some of the meat cut should be trimmed off. You can discard it as it has no nutritional value.

4. When you are finished trimming off the fat, the meat will be ready for slicing into strips.

The muscle bundles lay in the same direction as they attach to the skeleton although some may overlap and attach in different areas of the same bone. The length of the muscle fibers will have striations, or grains, that appear in a horizontal position in relation to the bone. Cutting the muscle length-wise is referred to as cutting "with the grain." Cutting across the muscle is referred to as cutting "against the grain."

## SHELF-STABLE JERKY

The USDA defines "shelf stable" as food that can be safely stored at room temperature or "on the shelf." These are nonperishable products that can include jerky, rice, pasta, flour, and sugar. Foods that can't be kept safe at room temperature, such as seafood, milk, and raw meats, are labeled "keep refrigerated."

A requirement to be shelf stable is that the perishable food must be treated by heat and/ or dried to destroy food-borne microorganisms that can cause illness or spoil food. Foods with this designation can be packaged in sterile, airtight containers. It is a standard rule that if not preserved in some manner, all foods will eventually spoil.

1. Using a mesh glove to firmly hold the meat while applying your knife can prevent accidental cuts to your hand.

2. Cutting with the grain of the meat will yield jerky that is easy to pull apart in strips.

3. Cutting against the grain of the meat will yield jerky that is easier to bite off in small pieces.

## SLICING STRIP JERKY

It is important to begin any cutting or slicing by using safe handling practices. Before the meat is taken from the refrigerator or freezer, clean all surfaces it will be exposed to. Wash your hands, knives, any mechanical slicer tray or blade, and all other surfaces the meat may come in contact with once is has been sliced, including meat trays. (Some slicers have trays attached alongside the blade to catch the meat slices as they are cut.) Be sure all parts of your slicer or your knife blade and handle have been thoroughly cleaned. Wash your cutting board or countertop and any bowls or utensils you will use with hot, soapy water, and then towel dry.

Whether using a mechanical slicer or knife, you can slice the meat either with the grain of the muscle fibers or against the grain. Slicing with the grain will give you slices that are easier to pull apart in strips because the tearing will follow the parallel meat fibers that lay lengthwise. Slicing the meat against the grain will produce a jerky that makes it somewhat easier to bite off a small piece of the jerky strip without having to rip it with your teeth. The slicing direction is more of a personal choice.

The slices should be no more than 1/4 inch (6 mm) in thickness but can be as long as the meat piece—as long as that length will fit in your dehydrator, of course. The thinner the strips are in thickness, the quicker they will dry.

If slicing with a knife, use a mesh glove on your free hand to avoid any injury from potential slippage of the blade. If you use a manual or electric mechanical

4. The muscle fiber direction will create striations. This will determine its grain when it's cut. Muscle fibers can run parallel in long lines or overlap to create cross fibers.

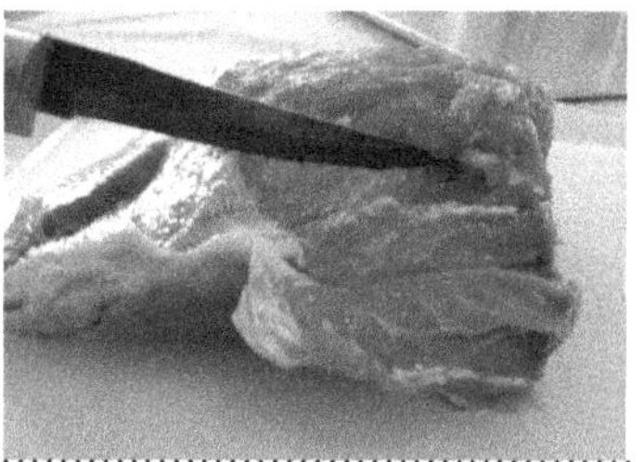

5. Cutting against the grain of the meat will yield jerky that is easier to bite off in small pieces.

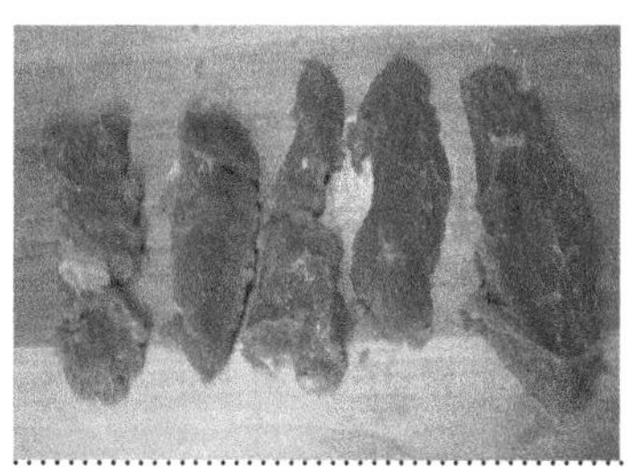

6. Your meat slices should be no thicker than 1/4 inch (6 mm). Strip width is not as much of a concern in the drying process as is thickness.

7. Keep a bowl of water available. Use it to dip your knife blade in periodically while slicing the meat. It will make the next cut easier.

8. A meat slicer can have three moving components: the blade, the handle to power it, and the meat tray, which moves back and forth.

9. This hand slicer model has an adjustment knob on the side that allows you to increase or decrease the thickness of the meat slices.

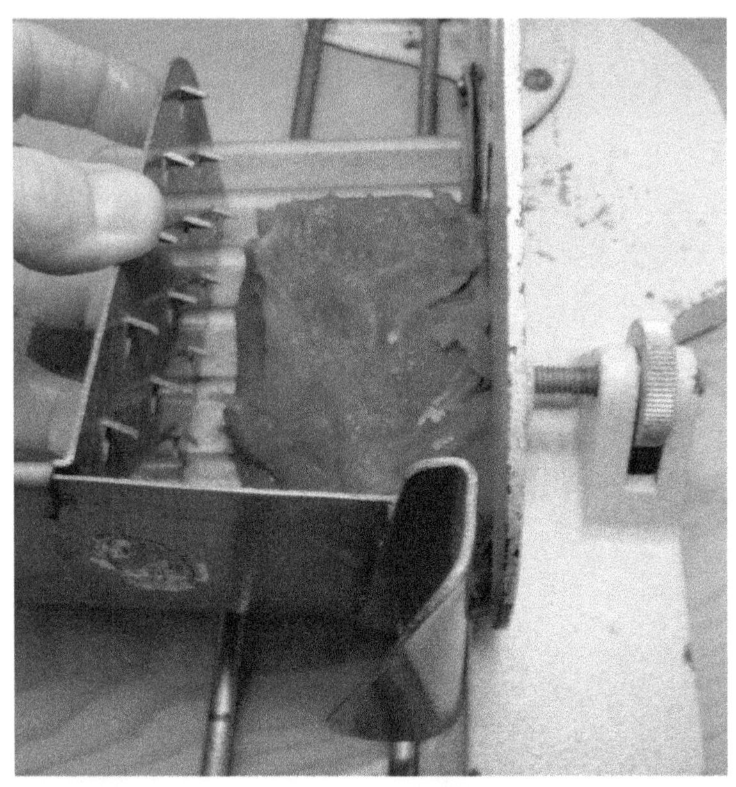

10. Place the meat cut into the tray and press the plate firmly against it. At the same time you push the tray and meat forward toward the blade, turn the handle to rotate the blade and make your first slice.

slicer, the process will be much quicker than a knife. Most slicers have adjustment options to vary the thickness of the slices. They work well for most meats except fish and fowl because of their texture and size of portions available for slicing.

Slicing meat while it is still frozen or partially frozen makes it easier to produce uniform width pieces than soft meat. That is because raw meat will tend to squash as you cut into it, which makes it more difficult to cut consistent slices.

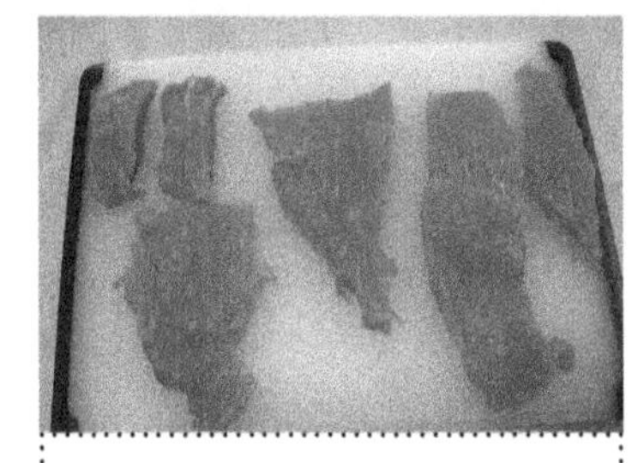

11. As each single piece is sliced, lay it aside to form rows in a clean pan or surface.

12. As the meat is sliced, it will separate and fall away from the blade.

Chunk, cube, or nugget jerky can be made from meat pieces too small to slice.

**NOTE:**
One option would be to cut one larger piece to use for your test. If it has been heated through to 160°F (71°C), you can be assured that all pieces smaller than it will have been as well.

### CUTTING CHUNK JERKY

While slicing will result in flat strips of jerky that look like bacon, you can also cut small chunks or nuggets of meat to make into jerky. Cutting chunk jerky is easier than cutting slices, but thicker chunks require more drying time because of the denseness of the meat, and the interior of the meat must reach 160°F (71°C). So time saved in one part of the process gets added in another. It really comes down to what you prefer to eat more than anything else.

One nice thing about chunk jerky is that you can use small or odd pieces of meat that are too small to be easily sliced into strips. This is often the case with upland game birds or waterfowl such as ducks and geese, and many small-game animals such as rabbits and squirrels.

## GROUND MEAT JERKY

Not all meat cut from a carcass will be the size you want for making jerky strips. Ground meat is a great way to use the scraps and small, oddly cut pieces made during the butchering process. Once ground, the meat can easily be formed into strips that are uniform in size and shape. There are other appeals to ground meat jerky as well. The resulting jerky is typically easier to chew than solid strips from one cut of meat. And some meats, such as the dark meat of ducks and geese, can be ground and mixed together with other meats with a milder flavor.

Due to the nature of the jerky-making process, any meat processed with a hand or power grinder should be kept cold. Also, if you are making ground meat jerky and purchasing the meat from a grocery store or market, avoid buying meat that has already been ground. This will typically have fat added to it in the store's grinding process, partly as filler and partly as weight. Because the meat has now been processed and placed in a package or case, it has greater potential to be affected by E. coli and other bacteria. As mentioned previously, the more meat is processed and cut up, the more surface area is exposed to potential contamination. Grinding the meat yourself will ensure you have control over the entire process.

There are two basic types of grinders, electric and hand-powered, and depending on your preference, either will work. An electric model will generally provide great speed with its blade. This can be both good and bad. Fast grinding speeds will process the meat more quickly but can also be a hazard if your fingers get too close. Some electric grinders may have speed settings that can increase or decrease the blade speed.

A hand-powered grinder's crank turns the internal worm gear to push the meat chunks into the cutting blade. These are slower than electric models and will require some physical effort. The more meat that is put into the chamber, the more resistance the crank will meet.

As with any jerky-making method, cleanliness and proper sanitation are key. Prior to using a hand-powered or electric grinder, disassemble all the parts possible and sterilize them in boiling water. Use this method of cleaning because the finer meat fibers and particles will be exposed to the inner surfaces of the grinder and plate. Sterilizing these surfaces before use will make sure there will be no contamination from the equipment. Be sure your electric grinder is not plugged in to an electrical outlet before you take it apart. Dry the parts and then put them back in place.

1. To slice meat for chunk jerky, begin with the smaller pieces. Slice them into about 1/4- to 1/2-inch (1.5 cm) squares. Smaller meat pieces can also be ground up for ground meat jerky.

2. An old-fashioned hand grinder can be attached to a table or counter. Newer electric models may be easier for you to use. Place a clean dish under the blades to catch the ground-up meat.

3. Meat chunks or strips can be dropped into the top throat of the grinder. Be sure to keep your fingers away from the moving parts, and use extra care when using an electric model.

4. Use a meat rod to push the strips or chunks further into the grinder. Never use your fingers.

**NOTE:**
It might go without saying, but make sure there are no bones in the meat that could get mixed in during the grinding process. Bones may stop an electric grinder or damage the worm gear, grinding plate, and blade. Also, any bones that get ground up may cause problems like chipped teeth, and they increase the chance of your jerky harboring bacteria.

It is more difficult to eliminate microorganisms that cause health problems in ground meat than with whole meat strips. It is vitally important that the oven, dehydrator, or smoker is capable of reaching temperatures of at least 200°F (95°C). This is to ensure that an internal meat temperature of 160°F (71°C) can be achieved to destroy any disease-causing bacteria present.

If you use an oven to dry the meat, preheat it to 200°F (95°C) before placing the meat inside. If using racks, place metal sheets underneath to collect any drippings. Heat the meat for 1 to 2 hours with the door slightly open. Maintain the heat until the meat strips have been dried to the point where they bend but do not break when tested. It will be more difficult to determine the internal temperature of strips because of their thinness. You must test the pieces, however, to be sure they have reached a temperature of at least 160°F (71°C).

## CUTTING THE MEAT

Let's go through the basic steps for grinding jerky.

1. Thoroughly wash your hands and any surface area that the meat may come into contact with before grinding. See pages 184 to 187 for more on cleaning and sanitation.

2. Begin by cutting the meat into small chunks that will fit into the top opening of your grinder, sometimes referred to as the throat.

3. After cutting all the meat, lay the pieces out on a clean surface and sprinkle them with your seasonings (see pages 206 to 210 for recommendations). This will help mix them into the meat as it is being ground.

4. Grind the chunks slowly. Use a low-speed setting on your electric model and do not overload the grinder.

5. Let the meat fall into a clean container as it moves out of the grinding tube.

6. Use a clean meat stomper or plastic rod to gently push any wayward chunks into the grinder's throat. DO NOT push with your fingers. Always keep your fingers away from the moving parts. Using a rod to push the pieces into the grinder will also protect your meat from possible contamination from your hands.

7. Make sure you have clean hands before handling the meat. Put the ground meat into the refrigerator and cover it until you work with it again.

8. With the meat safely put away, you can unplug your grinder, disassemble the parts, and take your time cleaning it. Wash all the grinder parts in hot, soapy water and rinse with hot water. Make sure to clean the worm gear, too.

9. After allowing the parts to dry, store them in a cool, dry space until their next use. There are food-grade silicone sprays that you can apply to the parts to prevent rust. The spray will be washed off at the time of your next use.

## SHAPING THE MEAT

Shaping the preprocessed jerky meat is required more with ground jerky than with strips or chunks, which for the most part retain their shape. Because ground meat is a soft mass, it will be more difficult to form into strips—but it is not an impossible task.

Ground meat can be formed either into thin strips or fashioned into small pieces or nuggets. Make sure to thoroughly wash your hands before handling the meat. You can either use clean rubber gloves, thin clear plastic food-handling gloves, or roll the meat with very clean bare hands:

1. Mix the cure, seasonings, and flavorings of your choice thoroughly with the ground meat before molding it into the desired shape.

2. Begin forming the meat into strips by placing it on a sheet of food grade freezer paper or waxed paper.

3. Place a second sheet of waxed or freezer paper over the top. Use a rolling pin to roll the meat out into the form of a patty about 1/4 inch (6 mm) thick.

4. Peel off the top paper and use a clean knife to slice the flat meat into strips about 1 inch (2.5 cm) wide. Make sure the knife isn't too sharp; you don't want to cut through the paper.

5. To avoid tearing the strips apart while transferring them to the jerky rack, flip the waxed or freezer paper onto the rack. Then, peel it off, leaving the strips behind.

6. Dry until an internal temperature reaches 160°F (71°C) to eliminate any disease-causing bacteria. Drying in an oven at a setting of 200°F to 275°F (95°C to 135°C) should help you attain the required internal temperature in one to two hours.

1. Before you shape the ground meat, add your seasonings and cure, and mix in a nonmetallic bowl or container.

2. You should accurately measure out the ingredients you add to the meat, especially the cure mixture. Always pay attention to the package directions if using a commercial mix. Add the cure and seasonings to cold water.

3. Then, pour over the ground meat and mix thoroughly.

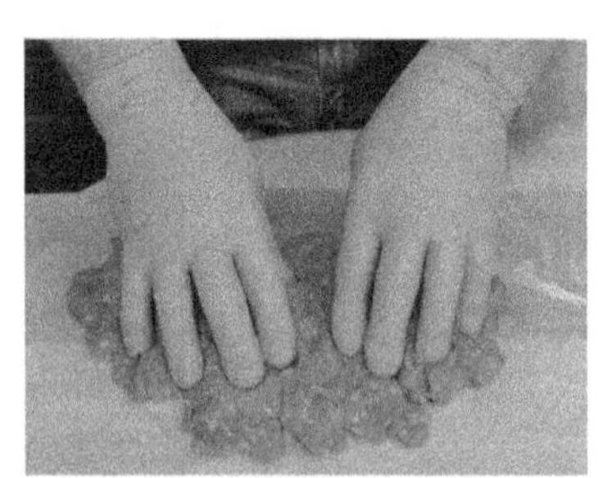

4. Ground meat jerky can be shaped into thin jerky-style strips with the use of waxed paper or the waxed side of freezer paper.

5. Spread and hand-flatten a small ball of ground meat onto a sheet of waxed paper. Then, place another sheet over the top.

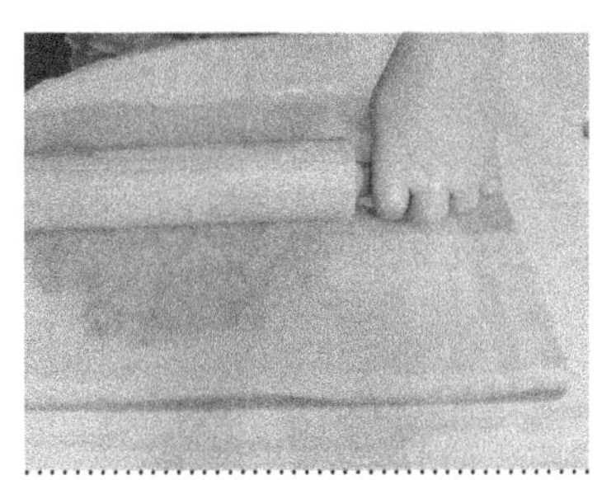

6. Use a rolling pin to roll the meat out flat to between a 1/8 and 1/4 inch (3 to 6 mm) thickness.

7. Peel off the top sheet of waxed paper. Use a clean table knife, without a sharp edge, to cut the meat into long even strips.

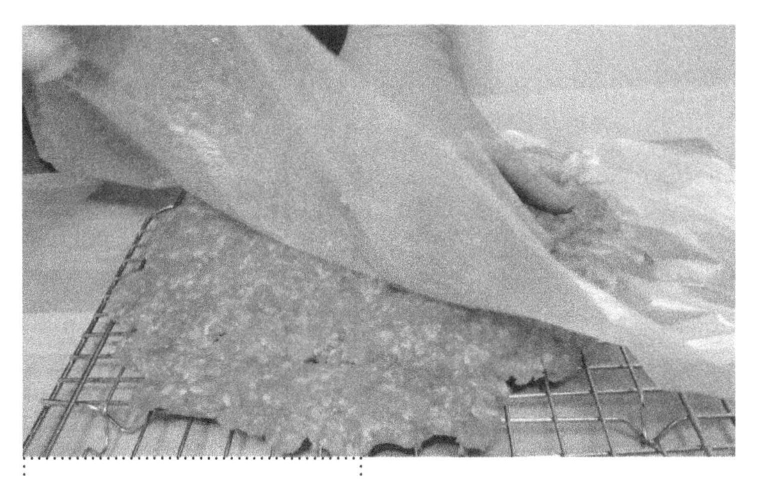

8. Place your meat on top of the oven rack or dehydrator tray. Then, lift the waxed paper with the cut strips on it and the rack.

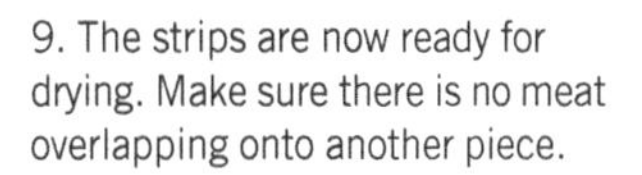

9. The strips are now ready for drying. Make sure there is no meat overlapping onto another piece.

If you find this method isn't for you, you can also try partially freezing the ground meat on plastic wrap by first flattening it out in a 1/4-inch (6 mm)-thick layer and putting it in the freezer. After hardening enough to handle easily but before completely frozen, cut it into strips. Peel off the plastic wrap and lay the pieces on the drying rack.

## USING EXTRUDERS

A meat extruder is a piece of equipment used to form meat into certain shapes, often flat or round, with an interchangeable tip. It resembles a caulk gun or a cookie-making gun. In this case, meat is packed into a hollow tube and pressure is applied against the meat by squeezing the handles together. This pressure then pushes an interior plate forward, and the meat is pushed out the opposite end in the shape formed by the tip.

While hand-held extruder models are easy to use, they may take some hand and arm strength to squeeze the handle, particularly at the start when the tube is completely full of meat. If that's a concern, you may consider buying a grinder that has an attachment for making meat strips or round sticks.

Here are the steps for using a hand extruder:

1. Make sure that all parts of a handheld extruder or a grinder-extruder model have been disassembled prior to use and thoroughly cleaned and dried. Wash your hands thoroughly before handling the meat.

2. To begin, wet your hands or gloves in cold water before scooping a cup of the cured and seasoned meat and placing it on a clean surface.

3. Pick up a portion of the meat and roll it between your hands to form it into a cylindrical shape that is small enough to slide into the tube. Then, add more small rolls until the tube is full. The meat will flow out more easily if it has been formed into thin rolls first rather than packing the tube full.

4. Place the plunger into the tube and secure the molding tip at the opposite end. Refer to the instructions that came with the extruder if necessary.

5. Spray your dehydrator tray, oven, or smoker rack with vegetable oil so the meat will not stick. Gently squeeze the handles of the extruder together and extrude the strips or sticks onto the tray or rack. Squeeze until the desired length is reached.

6. Once the tube is empty, you can remove the plunger and refill the tube.

7. Repeat the extruding steps until all of the meat you have prepared is used up. You are now ready to dry the jerky.

8. Thoroughly clean the tube, plunger, nozzle tips, and anything else used with hot, soapy water.

**NOTE:**

"Cure" can be used to describe either the actual process of turning raw meat into a preserved, safe, and edible meat or the commercial product pack, which contains the essential ingredients that creates the transformation.

1. Any leftover ground meat can be shaped into sticks by hand and put into an extruder gun. A second method of using leftover meat is to form it into a 1-inch (2.5 cm)-thick block or slab, wrap it in plastic, and then freeze it. Then, remove the wrap and slice into 1/4-inch (6 mm)-thick strips.

2. Fashion the ground meat into small sticks by rolling a handful between your fingers or hands. Wet your hands or gloves with cold water to keep the meat from sticking.

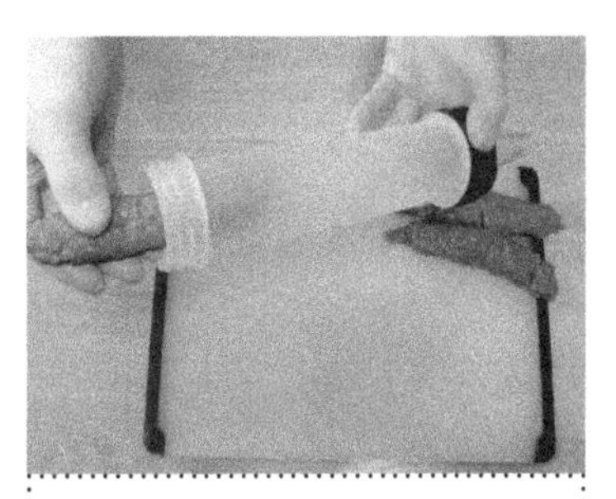

3. Slide the meat rolls into the extruder tube and reassemble the handle and plunger.

4. Gently squeeze the handle to push the ground meat out the tip and onto a rack or screen. The resulting extruded ground-meat jerky is ready to place in the dehydrator or oven.

5. Space the ground meat jerky in the dehydrator to encourage even airflow. You may want to rotate the racks from top to bottom during the drying process.

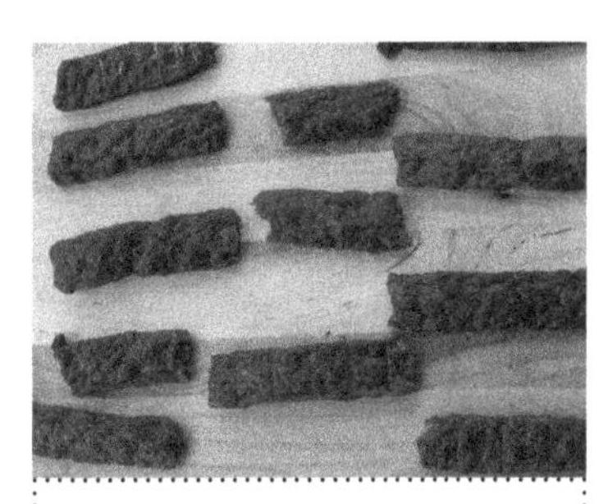

6. Ground meat jerky will have an elongated shape if you use an extruder. Whether strips or sticks, allow them to cool in the refrigerator after you remove them from the dehydrator.

Cures and seasonings add flavor to the jerky and help with preservation. These can be added to the ground meat before shaping. Be sure to thoroughly mix the cure and seasonings with the meat. If you use a commercial cure mix, follow the directions exactly regarding the amounts and mixing procedure. If you create your own cure, be sure to use precise amounts of all ingredients called for in your cure recipe and calculate precisely.

## CURING AND MARINADES

Curing is generally done using either a liquid marinade into which the meat is submerged or a dry rub, which is used to coat the meat's exterior. Typically, the meat must sit for a certain length of time before drying in order for the cure to work through the exterior meat surface and into the interior. This is accomplished by osmosis as the salt slowly moves through the cell membranes. Generally speaking, the larger the piece of meat being cured, the longer it will take to complete the curing process.

Marinating meat or vegetables involves soaking foods in a usually acidic liquid that often contains flavorings, seasonings, and salts. Depending on the food used and personal preferences, this process may last several minutes, hours, or days.

The acids in marinades help to break down the surface tissues, which some advocates believe allows more of the seasoned liquids to be absorbed inside, resulting in a more flavorful product. Marinades, however, react more to the surface of the meat rather than penetrating inside. Meat protein contains about 75 percent water and thus has little room to absorb more moisture.

Marinating is not tenderizing, which is a process that causes proteins in meat to soften and make it more tender to the chew. A good marinade must balance the acid, oil, and spice in order for the meat to not become too acidic, which can produce harsh or off flavors.

Always marinate meat in your refrigerator. Never marinate at room temperature, because this will put the meat in dangerous temperatures where bacteria multiplies quickly (the danger zone is between 40°F and 140°F [4°C and 60°C]). Never use a recipe that calls for marinating at room temperature.

**NOTE:**
Mix only the meat you can process at one time, and use it as soon as it's mixed. Do not store meat in tubes or nozzle tips overnight as it will be very difficult to push out after it has cooled to refrigerator temperatures.

The drying time is not significantly affected by marinating the meat. The temperature at which the meat is dried is more important than the length of time spent drying, although these two dynamics have required levels to produce a safe product (more on this on pages 211 and 214 to 216).

Ground meat used for jerky making is typically not placed in a marinade because it can become too soft and unmanageable when formed into strips or sticks. Liquids in small amounts and dry ingredients can be successfully mixed into ground meat without it losing the firm consistency necessary for handling.

## SALT

Salt is the essential ingredient in the curing process. It draws moisture and blood from the muscle cells while entering the cells by osmosis. This process distributes the salt through the tissue and induces partial drying. Salt also inhibits the action and function of certain bacteria and enzymes.

This is why salt is one of the most common ingredients used in the production of jerky. It is central to all curing mixtures. Salt is hygroscopic, which means it absorbs moisture from its surrounding environment. This is its preservative action.

Choose a good-quality, food-grade salt for your curing process, such as a coarse pickling or canning salt. These readily dissolve in liquids, which make them easy to use. Using a non-iodized salt in a marinade or brine is also best as it prevents potential chemical reactions if you're using heat.

Salt makes up the bulk of any curing mixture not only because it is a good preservative: it also provides a desirable flavor. Curing salt (sodium nitrite) can be used in homemade jerky, although it is not required. Commercial cures are readily available. These typically contain some form of salt as the main preserving agent but can also contain sugar, sodium nitrate, sodium nitrite, and propylene glycol (which helps to keep the mixture uniform).

There are also store-bought mixes available that contain all ingredients and spices in one packet, such as Morton Tender Quick, Hi Mountain Jerky Cure and Seasoning, Presto Seasonings, and Nesco. Using these will help speed up your jerky production. These mixes can serve as a base for exploring your own variations as long as the rates and ratios used coincide with the amount of meat being used.

Failing to maintain the proper level of curing ingredients in relation to the meat weight volume will lead to underprocessing. The sugar or pepper levels, for instance, can be increased to make a sweeter or more flavorful final product, but neither will affect the curing of the jerky.

Curing can be accomplished by using one of two methods: marinating the meat in liquid or coating it with a dry rub. Typically, the meat must be in contact with the marinade or rub a length of time before drying in order for the cure to work through the muscle fibers.

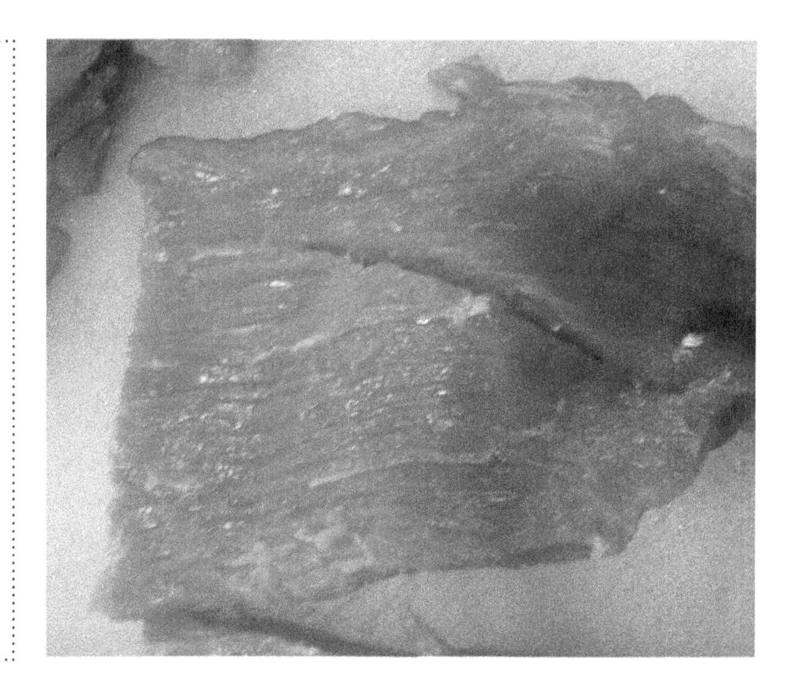

## MARINADES

Marinades are liquid brines or pickles in which meat or fish are soaked to enrich flavor prior to drying. A wide range of flavors can be created and may only be limited by your imagination. Some ingredients in marinades, such as salt and sodium nitrate, cure the meat and help preserve it. Other ingredients, such as vinegar and soy sauce, will help enhance the flavor, texture, and even appearance of the meat. Generally speaking, the amount of marinade needed is dependent upon the amount of meat being processed. If you are allowing the meat to marinate overnight, make sure all pieces are completely submerged in the liquid. On the other hand, you will need less marinade if you are just dipping the meat pieces into it and then laying them on a rack.

1. A liquid marinade is often used for jerky. Thoroughly mix all ingredients in a nonmetallic container.

2. Submerge the meat strips into the marinade until both sides are completely covered. Cover the container and let it sit in your refrigerator for 8 to 10 hours, or overnight.

## FLAVORINGS

Depending on your tastes, there are several flavoring agents that can be used when curing jerky, including herbs and spices, sugar, liquid smoke, and oil.

*Herbs and spices*: Whether added whole, crushed, ground, pureed, fresh, or dried, there are many herbs and spices, as mentioned earlier, that can be used to add flavors in jerky. You can include such herbs as basil, oregano, sage, mint, dill, rosemary, thyme, chives, parsley, or savory. Spices such as cayenne, chili powder, curry, garlic, ginger, allspice, mustard, nutmeg, and horseradish are only a few that are available, depending on your tastes. One of the most common spices used in jerky making is pepper. Black pepper is most used because it adds flavor and color.

*Sugar*: Sugar is an important ingredient because it helps moderate the intensity of the salt flavor. This moderating effect helps reduce the perception of saltiness.

*Liquid smoke*: Liquid smoke is a popular flavoring for jerky, as discussed on page 159. However, if too much is added, it can overpower the meat flavor and even create a harsh, bitter flavor: this is an instance when using more is not better. Liquid smoke can be used with food dehydrators and is the only way to add a smoky flavor in a dehydrator.

*Oils*: Some oils used in jerky making include those derived from olives, sesame seeds, and basil. Oils can add flavor but will not dry on the jerky. They can help add texture but will need to be patted off the jerky with paper towels during the drying process.

*Other ingredients*: Marinades can also include onions, fruits, or vegetables. Juices from lemons, oranges, apples, or tomatoes can be used, too, along with various kinds of peppers. Crushing, mincing, and mashing these ingredients will create more surface area to transfer their flavor to the liquid.

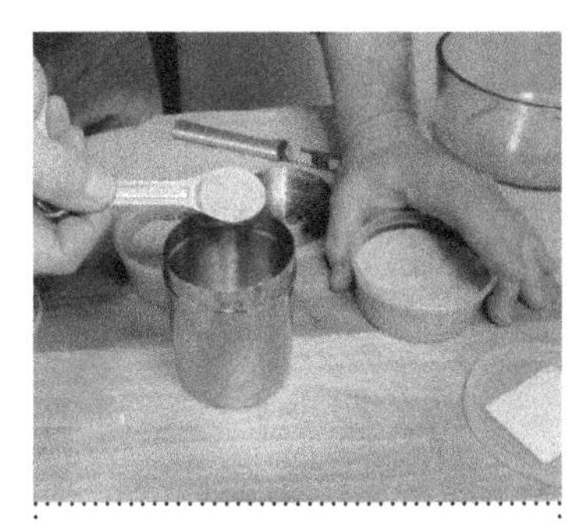

1. A dry rub cure can be sprinkled or rubbed onto the jerky meat. Mix the ingredients thoroughly before applying them to the meat.

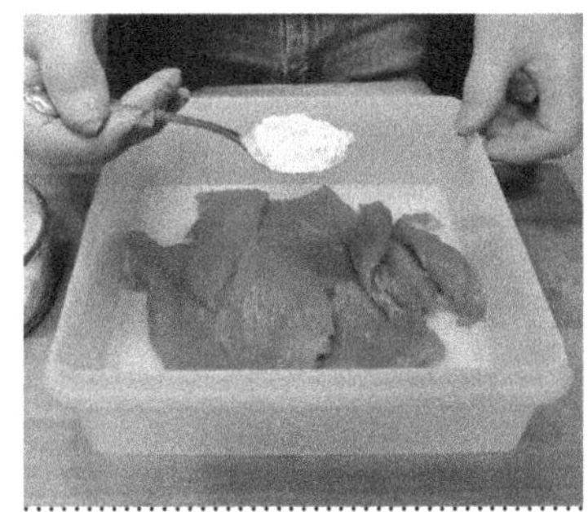

2. Place the meat sttrips in a nonmetallic container and sprinkle the dry rub on the top of the meat.

3. Next, roll the meat strips in the dry rub.

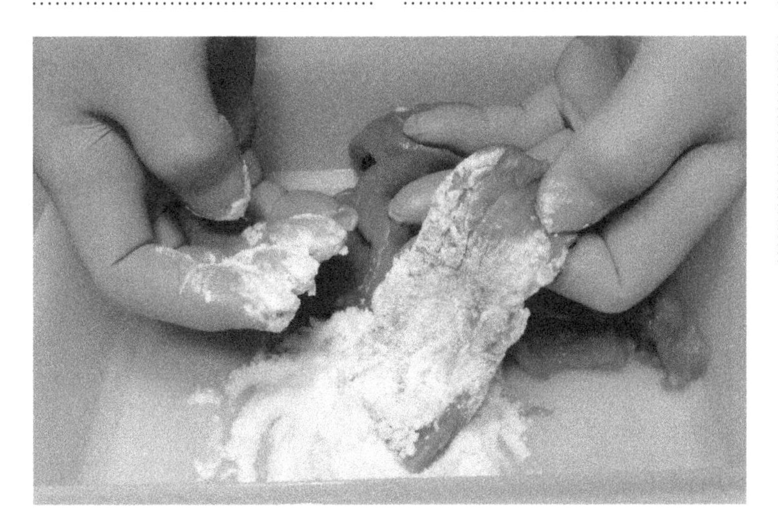

4. Coat all sides thoroughly. For most cure and seasoning mixes to be used properly and safely, you must know the exact amount of meat being used.

*(continued)*

5. Place the rubbed strips in a glass or plastic container or in a resealable plastic bag to cure.

6. Different marinades can be used with ground meat jerky. Thoroughly mix the ingredients in a nonmetallic container.

7. Then, pour evenly over the ground meat.

8. Mix the ground meat and marinade thoroughly before fashioning into strips.

9. These pieces of ground meat bison jerky have been infused with cranberries and black pepper.

# FOUR WAYS TO MAKE JERKY

Given sufficiently low humidity and enough sun, thin slices of meat will dry in the open air. While this primitive method may have worked for Native Americans and pioneers, it is not recommended today because it can foster bacterial growth and expose the meat to insect or animal contamination and spoilage. Instead, you will need to use a dehydrator or a smoker, sometimes in combination with an oven. The equipment you use does not need to be expensive or fancy, but it must be reliable, especially with respect to temperature control.

It was long believed that meats for jerky only needed to reach 145°F (63°C) to ensure the safety of the finished product. Some years ago, however, several incidents of food-borne illnesses linked to commercial and homemade jerky spurred research by food scientists into the processes needed to inactivate bacterial pathogens in the meat. As a result of that research, the USDA now recommends that to make jerky safely you should heat the meat to an internal temperature of 160°F (71°C) (poultry to 165°F [74°C]) before drying. Fish can still be heated to 145°F (63°C) and be used safely.

Here, we'll focus on four ways to make jerky:

1. High-Temperature Dehydrator (pages 214 to 215)
2. Oven and Low-Temperature Dehydrator (pages 215 to 216)
3. Smoker and Low-Temperature Dehydrator (page 218)
4. Smoker (page 219)

## DEHYDRATORS

There are two main types of home dehydrators available today. The first is the stacking dehydrator. Sometimes called "round" dehydrators (although some are actually rectangular in shape), the stacking dehydrator is composed of an electrical heating element and a fan, which are located either at the top or the base of the unit, and which work together to circulate heated air around vertically stacked trays of food. The trays are usually perforated with a large hole in the center of each tray to enable the heated air to circulate properly through the stacked trays. The warm air circulating around the trays dries the food. Stacking dehydrators tend to be fairly compact and relatively inexpensive, and some models allow their owners to expand drying capacity simply by adding additional trays that can be purchased from the manufacturer.

Your home oven can be used together with a dehydrator to safely make jerky. See pages 215 to 216 for instructions.

Jerky strips can be suspended from oven racks for drying. Use a toothpick and slip it through the top of each strip and position them on the racks so they do not touch each other. Be sure to cover the bottom of the oven with tin foil before preheating it and adding the meat. The foil will catch the drippings as the meat is drying.

To release some of the moisture from the oven, use a wooden spoon to crack the door. Continue to monitor the temperature during the drying process so the appropriate temperature is maintained. Do not leave it unattended.

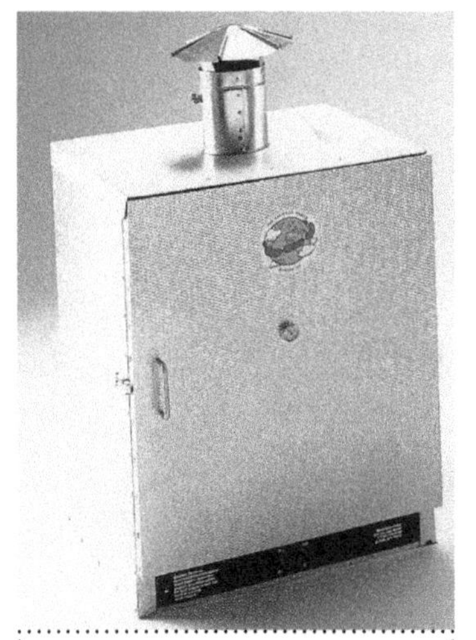

Insulated variable-temperature smokers can provide reliable temperature control while taking up little space.

**NOTE:**
There are a few dehydrators that have no fan, but instead rely entirely on convection to heat and dry the food. They're quiet and do a good job on some kinds of foods, but much longer drying times makes them less than ideal for making jerky.

The main disadvantage of most stacking dehydrators is inherent to their design. As the heated air moves through the multiple trays of food, it cools, so food in the trays farthest from the heating element dries more slowly than food in the trays closer to the heat source. The trays need to be rotated during the drying process to get all the food to dehydrate at roughly the same rate, and the more trays you have, the more diligently you have to rotate them. It's noteworthy that some popular Nesco stacking dehydrators solve this problem with a technology that pushes the heated air through an exterior pressure chamber and then horizontally across the food, substantially reducing the need to rotate trays.

Stacking dehydrators that have the heating element and fan at the bottom of the appliance are often the least expensive dehydrators. If moisture and crumbs fall from the trays into the heating element and fan, however, these dehydrators can be hard to clean—and their durability may suffer.

Some popular models of stacking dehydrators are the Nesco FD-80A, the Waring Pro, the Presto 06301, the L'Equip 528, and the Nesco Snackmaster Pro.

Avoid using a dehydrator that has an exposed heat coil at its base. Any fat from drying meats that drips onto this coil can create a fire hazard. If yours has one, be sure to cover the coils with aluminum foil so any drippings don't reach the coils.

The second type of home use dehydrator is the side-fan dehydrator, sometimes called a "square" dehydrator. These have a heating element and a fan mounted on the back side of the device, blowing warm air across multiple trays that are made to slide in and out of a door on the opposite side of the unit. This arrangement helps ensure fast and even drying of the food, as well as the durability of the appliance. And it's easy to check the progress of the drying process, which can be troublesome with a stacking dehydrator, which requires that you unstack the trays to see the food inside. Side-fan dehydrators, however, are generally more expensive than stacking dehydrators, and they take up more counter space. Also, the capacity of a side-fan dehydrator can't be increased by adding additional trays. Raw-food enthusiasts and others who use their dehydrators frequently often prefer side-fan models. Some popular side-fan dehydrators include the Excalibur 3500B and 3926TB.

Both kinds of dehydrators can be used for making jerky, but the process you'll use depends on a key variable: the temperature that can be reached by your dehydrator. Many inexpensive dehydrators have only one temperature setting, which is set by the manufacturer. Designed primarily for drying fruits, nuts, and vegetables, these dehydrators usually will heat to around 140°F (60°C), which isn't high enough to ensure food safety in meat jerky. If you have one of these single-setting dehydrators or a variable-temperature model that won't reach 160°F (71°C), you'll need to precook the meat to an internal temperature of 160°F (71°C) (165°F [74°C] for poultry) before putting it in your dehydrator to dry. For our purposes, any dehydrator that can reach 160°F (71°C) (or 165°F [74°C] if you intend to make poultry jerky) is a "high-temperature" dehydrator, while dehydrators that won't reach that temperature are "low-temperature" dehydrators.

So, before drying meat in a low-temperature dehydrator, you will first cook it in your oven to an internal temperature of 160°F (71°C) (165°F [74°C] for poultry). After heating your meat to 160°F (71°C), you'll move it to your dehydrator and maintain a constant dehydration temperature of 130°F to 140°F (54°C to 60°C) for 4 to 6 hours or until it's done. This is important because the process must be fast enough to dry food before it spoils, and it must remove enough water to prevent microorganisms from flourishing. If the proper temperature is maintained, this two-step process will yield a safe finished product. Of course, if you have a high-temperature dehydrator that will heat the food to 160°F (71°C) (or 165°F [74°C] for poultry), you don't have to precook your jerky in the oven.

Meat jerky made in a high-temperature dehydrator is a little different from that made by the two-step oven-and-dehydrator process. Jerky that's first cooked in the oven tends to be a little lighter in color and a little more tender than jerky prepared entirely in a high-temperature dehydrator. Some people prefer the jerky made by the two-step process, but jerky dried entirely in a high-temperature dehydrator looks and tastes more like what most people think of as traditional jerky.

Depending on a number of variables, such as meat thickness, amount of brine or liquid smoke used, and the dehydrator itself, you will need anywhere from 4 to 24 hours to complete the drying process.

The importance of proper drying techniques and temperatures cannot be overstated. Proper drying reduces the external and internal moisture content of the meat to levels that inhibit bacterial activity and growth.

## CALIBRATING THERMOMETERS

Before you move on to the four methods for making jerky, make sure your thermometers are calibrated, as discussed earlier. This may seem tedious, but it is important. Make certain the calibration of your dial-stem thermometer is correct. Simply put, you are checking the accuracy of the thermometer that will check your dehydrator.

**NOTE:**

Fats and oils typically do not dry. This is due to their composition and the melting points needed to disperse their molecules. This is one reason for cutting off or trimming as much fat as possible when you are preparing meat and fish for making jerky: it reduces the likelihood that fat will become rancid and spoil your jerky after drying. In wild game, removing fat also helps reduce the gamey flavor. If using fatty meats in a dehydrator or oven, use a paper towel to pat off any fat beads that form on the meat strips during the heating process.

## CALIBRATING YOUR EQUIPMENT

Having assured yourself that your thermometer will give you an accurate reading, you are now ready to determine your dehydrator or smoker's drying temperature. Do not test the temperature when the device has meat in it. You will get an inaccurate temperature reading if you do so because of the evaporative cooling that occurs as the meat loses moisture.

For ovens and side-fan dehydrators, you can place a dial-stem thermometer inside the unit and close the door. If your unit is a stacking dehydrator, insert the thermometer between two trays so that the dial sticks out between them.

Next, turn the dehydrator on and to its maximum setting. Once the unit has run for a minimum of 10 minutes, the temperature should stabilize and you can record it. Recheck after another 10 minutes to be certain it has maintained an internal temperature of at least 145°F to 155°F (63°C to 68°C). If it does not maintain this temperature, you will need to review the manufacturer's warranty or determine the cause of your unit's failure.

Be aware that most dehydrators equipped with temperature controls and thermostats will cycle around an average temperature at any given setting. For example, at a setting of 155°F (68°C), a dehydrator may cool to 150°F (66°C) and then heat up to 160°F (71°C), repeating this cycle every few minutes. That's not necessarily a fault. The popular Excalibur side-fan dehydrators are deliberately designed to do this, as the manufacturer believes that cycling around an average temperature creates a better finished product. Dehydrators that operate with a single, factory-set temperature will tend to maintain a more stable temperature if they are operating correctly.

The University of Wisconsin–Extension actually recommends that you do not use dehydrators with preset factory temperature settings that can't be controlled, as they do not reliably produce a safe product. You can, however, check your dehydrator's ability to reach and hold target temperatures by using a reliable and accurate dial-stem thermometer, placing it inside the dehydrator and checking it several times over 10 minutes to ensure that it can be trusted.

# MAKING JERKY WITH A DEHYDRATOR

Although it's possible to make jerky in a conventional oven, I don't recommend it because drying times are long and temperature control is difficult, as you have to keep the oven door partly open to promote air circulation. A convection oven, however, promotes air circulation with a fan and can be used much like a dehydrator to make good jerky. It's best not to dry too much meat at once, however, as most convection ovens do not dry foods as efficiently as dehydrators do. If you use a convection oven to make jerky, use the instructions for making jerky with a high-temperature dehydrator (Method 1).

## METHOD 1: HIGH-TEMPERATURE DEHYDRATOR

You will need:

- Food dehydrator that can reach temperatures of 160°F (71°C) or higher
- 2 pounds (1 kg) of meat and a marinade
- Equipment used for slicing meat strips
- Drying racks or trays

## INSTRUCTIONS

1. Thoroughly wash your hands, countertops, knives, slicers, and any other pieces of equipment you will use.
2. Slice the meat into strips at about 1/4 inch (6 mm) thick and prepare per your recipe. Most recipes will require marinating for at least 2 hours or overnight before proceeding.
3. Set out the drying racks or trays, remove the meat strips from the marinade, and place the meat strips close together but not so close that they touch or overlap each other. Place the trays or racks in the dehydrator.

4. Preheat a reliable high-temperature dehydrator to 160°F (71°C) or slightly higher. Then, put in the meat for 4 to 6 hours or until it reaches an internal temperature of 160°F (71°C) (poultry needs to reach 165°F [74°C] internal temperature). If you maintain the high temperature, you should not need to lower it to finish the drying process. However, monitor the meat strips' progress so that they do not become too dry.

5. You should begin checking the jerky after 3 hours from the start of the drying process. To test the jerky for doneness, first pat off any beads of oil or fat with a paper towel. The jerky can be considered finished if it cracks when bent over on itself but doesn't break clean through. There is a fine line between underdone and overdone jerky. If it breaks, it has been dried too much. This isn't a disaster—it's still edible. With any new equipment that you use, some experimentation may be necessary to reach a result that is satisfactory to your tastes.

6. After allowing the jerky to cool to room temperature, store it in a clean plastic container with a tight-fitting lid or in a resealable plastic bag, and place the container in the refrigerator. If you notice any condensation forming on the inside of either, the jerky should be returned to the dehydrator and dried a little longer.

Jerky that is properly dried should bend a little without breaking but not be so soft that it bends completely.

## METHOD 2: OVEN AND LOW-TEMPERATURE DEHYDRATOR

You will need:

- Oven
- Food dehydrator
- 2 pounds (1 kg) of meat and a marinade
- Equipment used for slicing meat strips
- Drying racks or trays

## INSTRUCTIONS

1. Thoroughly wash your hands, countertops, knives, slicers, and any other pieces of equipment you will use.

2. Slice the meat into strips at about 1/4 inch (6 mm) thick and prepare per your recipe. Most recipes will require marinating f or at least 2 hours or overnight before proceeding.

3. Preheat your oven to 145°F to 155°F (63°C to 68°C). Set out the drying racks or trays, remove the meat strips from the marinade, and place the meat strips close together but not so close that they touch or overlap each other.

4. Once the oven is preheated, place the trays or racks in the oven. Bake the meat for approximately 4 hours. Increase the heat to 275°F (140°C, or gas mark 1) and continue to bake until the meat reaches an internal temperature of 160°F (71°C) (poultry needs to reach 165°F [74°C] internal temperature).

5. Once the meat reaches an internal temperature of 160°F (71°C), you can use a low-temperature dehydrator to finish the drying. Transfer your meat to a dehydrator and maintain a constant dehydrator temperature of 130°F to 140°F (54°C to 60°C) while you continue drying.

6. You should begin checking the jerky after 3 hours from the start of the drying process to make sure that is doesn't dry past the point where it's enjoyable to eat. To test the jerky for doneness, first pat off any beads of oil or fat with a paper towel. The jerky can be considered finished drying if it cracks—but doesn't break—when it's bent over on itself. There is a fine line between underdone and overdone jerky. If it breaks, it has been overheated. This is not a disaster—it will still be edible—but more of an inconvenience. With any new equipment that you use, some experimentation may be necessary to reach a result that is satisfactory to your tastes.

7. After allowing the jerky to cool to room temperature, store it in a clean plastic container with a tight-fitting lid or in a resealable plastic bag, and place the container in the refrigerator. If you notice any condensation forming on the inside of either, the jerky should be returned to the dehydrator or oven and dried a little longer.

# MAKING JERKY WITH A SMOKER

Smoking jerky meat can be done for one of two reasons: it can add smoky flavor and an attractive color to your jerky and/or it can dehydrate your jerky completely—similar to using a high-temperature food dehydrator.

## LIQUID SMOKE

If you're after a smoky flavor, you can, of course, bypass the smoker and use liquid smoke as discussed earlier. It can be added to a marinade or dry rub and will adhere to the surface of the meat before it's dried. This coating will provide a smoky flavor. It should be used in moderation and in keeping with recipe recommendations to avoid off flavors. See page 159 for more.

## SMOKERS

If you just want to add a smoky flavor to your meat before using a dehydrator, you can use just about any model—even a home grill with a smoke box or smoking attachment will work just fine if the proper temperatures can be reached and maintained for extended periods. Traditional smokers burn wood both to create smoke and to cook the meat. With electric smokers, you will need to add wood chips or pellets to a heating chamber to produce smoke.

Store your dried jerky in glass jars at cool or room temperatures and away from sunlight and humidity, or in the refrigerator.

Using a smoker to make jerky will require some trial and error with your particular model. It's safe to say, however, that if you're already quite comfortable with it—if you know how to get it to hit key temperatures and keep it there—you'll be ahead of the game.

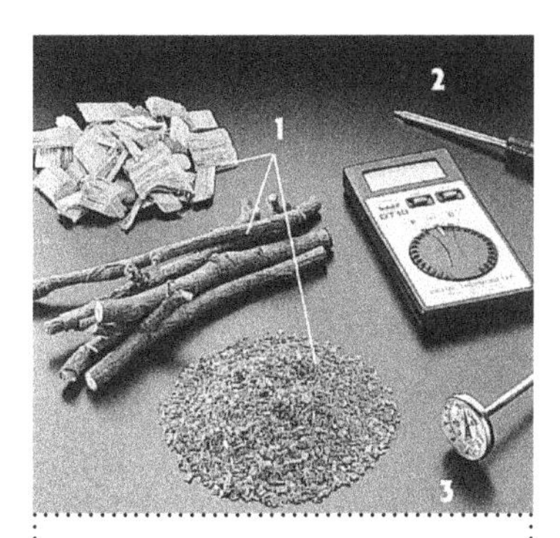


To produce the smoke needed, various woods (1) such as hickory, cherry, apple, and mesquite are used in home smokers. The smoke can be made from chips, tree trimmings, or sawdust. A digital thermometer (2) monitors the internal temperature without having to open the smoker, causing heat loss. An instant-read thermometer (3) monitors the temperature through a vent hole or a hole specifically designed for a thermometer.

## STORING JERKY

Dried jerky can be safely stored for one to two months at room temperature and in the freezer for up to six months.

Do not consume any jerky stored in a container that shows any signs of mold. Dispose of it immediately and rinse any affected container with scalding water first. Then, use a sanitizer to thoroughly clean the container and rinse again with scalding water. Be careful when handling any affected pieces by using rubber or plastic gloves and thoroughly wash the gloves after handling moldy pieces or containers. Make certain the affected pieces or containers do not come into contact with counter surfaces or other foods to be consumed. Be sure that when you dispose of any moldy jerky pieces that children or pets are not able to come into contact with them.

No matter your level of familiarity with your smoker, it will require attention and patience to control several variables when making jerky. For example, to provide a good ventilation of the smoke while still maintaining the proper heat, you need to make sure the smoker has proper airflow. Also, while it might seem like a contradiction, it may be necessary to maintain a degree of humidity inside the smoker through the dehydrating process to prevent the meat from getting too dry or too smoky.

Some precautions may be necessary. Typical recommendations suggest smoking the jerky at 200°F (95°C) for 1.5 to 2 hours with the smoke on, but a little longer if the smoker does not reach that temperature. Don't smoke the meat for more than 3 hours because too much smoke can produce a bitter taste.

The most commonly available home smokers that work for making jerky include vertical electric water smokers, insulated variable-temperature smokers, electric smokers, and stove-top smokers, as discussed earlier.

Depending on the model used, one concern about smokers is the potential loss of heat caused by opening a door to add water to a pan or to replenish wood chips or pellets. Models are available that have external wood chip or pellet loaders so the unit doesn't need to be opened. A smoker like the Masterbuilt Electric Smoker has a tray that can be pulled out, have fresh chips or pellets added to it, and be pushed back into the unit without opening the door. (This is an outdoor model not meant for inside use.)

## METHOD 3: SMOKER AND LOW-TEMPERATURE DEHYDRATOR

You will need:

- Smoker (outdoor model)
- Food dehydrator
- 2 pounds (1 kg) of meat and a marinade
- Equipment used for slicing meat strips
- Drying racks or trays

## INSTRUCTIONS

1. Thoroughly wash your hands, countertops, knives, slicers, and any other pieces of equipment you will use.

2. Slice the meat into strips at about 1/4 inch (6 mm) thick and prepare per your recipe. Some recipes may require marinating for at least 2 hours or overnight before proceeding.

3. Preheat your smoker to 200°F (95°C). Add chips or pellets to the burn chamber to begin smoking. Use the smoker vents to stabilize the temperature between 165 and 175°F (79°C).

4. Set out the drying racks or trays. Remove the meat strips from the refrigerator and the marinade. Pat the strips dry and place them close together—but not so close that they touch or overlap each other.

5. Place the rack or racks with your meat in the smoker. Insert a temperature probe into the thickest piece of meat.

6. Smoke the meat for 4 to 6 hours or to an internal temperature of 160°F (71°C) (or 165°F [74°C] for poultry). The amount of time smoke is added to the heating chamber will depend on personal preference, but 3 hours should be sufficient. If you prefer a heavy smoke, you can allow more time. You may have to experiment with the smoking process to determine the amount that suits your tastes. Add water to the pan if needed—although you are dehydrating the meat, it may be necessary to maintain humidity inside the smoker to prevent it from getting too dry or too smoky.

7. Once the meat reaches an internal temperature of 160°F (71°C), you can use a low-temperature dehydrator to finish the drying. Transfer the meat to the dehydrator and maintain a constant dehydrator temperature of 130°F to 140°F (54°C to 60°C) while you continue drying.

8. Test the jerky for doneness after 4 hours. First pat off any beads of oil or fat with a paper towel. The jerky is finished if you can bend it over on itself and it cracks but doesn't break. There is a fine line between underdone and overdone jerky. If it breaks, it has been dried too much. This is not a disaster—it will still be edible—but more of an inconvenience. With any new equipment that you use, some experimentation may be necessary to reach a result that is satisfactory to your tastes.

9. After allowing the jerky to cool to room temperature, store it in a clean plastic container with a tight-fitting lid or in a resealable plastic bag, and place the container in the refrigerator. If you notice any condensation forming on the inside of either, the jerky should be returned to the dehydrator and dried a little longer.

## METHOD 4: SMOKER

You will need:

- Smoker (electrically powered)
- 2 pounds (1 kg) of meat and a marinade
- Equipment used for slicing meat strips
- Drying racks or trays

## INSTRUCTIONS

1. Thoroughly wash your hands, countertops, knives, slicers, and any other pieces of equipment you will use.

2. Slice the meat into strips at about 1/4 inch (6 mm) thick and prepare per your recipe. Most recipes used in this book require marinating for at least 2 hours or overnight before proceeding.

3. Preheat your empty smoker to 200°F (95°C) and hold at 175°F to 180°F (79°C to 82°C) for 15 to 20 minutes. Add water to the pan as well as seasonings (if desired).

4. Set out the drying racks or trays. Remove the meat strips from the refrigerator and the marinade. Pat the strips dry and place them on the racks so that they're close together, but not so close that they touch or overlap each other.

5. Place the rack or racks with your meat in the smoker. Insert a temperature probe into the thickest piece of meat.

6. Heat the meat for 4 to 6 hours or to an internal meat temperature of 160°F (71°C). You may wish to rotate the racks after 2 hours to ensure even heating.

7. After the internal temperature reaches 160°F (71°C) (165°F [74°C] for poultry), begin adding smoke to the heating chamber with wood pellets or chips in the chamber designed to hold them. The amount of time smoke is added to the heating chamber will depend on personal preference, but 3 hours should be sufficient. If you prefer a heavy smoke, you can allow more time. You may have to experiment with the smoking process to determine the amount that suits your tastes.

8. Test the jerky for doneness after 4 hours. First pat off any beads of oil or fat with a paper towel. The jerky is considered finished if you can bend it over on itself and it cracks but doesn't break. There is a fine line between underdone and overdone jerky. If it breaks, it has been overheated. This is not a casualty—it will still be edible—but more of an inconvenience. With any new equipment that you use, some experimentation may be necessary to reach a result that is satisfactory to your tastes.

9. Remove the racks from the smoker. After allowing the jerky to cool to room temperature, store it in a clean plastic container with a tight-fitting lid or in a resealable plastic bag, and place the container in the refrigerator. If you notice any condensation forming on the inside of either, the jerky should be returned to the smoker and dried a little longer.

# GLOSSARY

**Aging:** The time process that causes a maturing or ripening of meat enzymes to increase flavor and tenderize

**Aitchbone:** The rump bone

**Anterior to:** Toward the front of the carcass

**Antioxidant:** A substance that slows down the oxidation of oils and deterioration

**Backstrap:** Connective tissue composed of elastin that is found in the neck. In most animals the backstrap is inedible

**Blade meat:** An inedible, yellowish-colored connective tissue composed of elastin running from the neck through the rib region of beef, veal, and lambs

**Butterfly:** To split steaks, chops, cutlets, and roasts in half, leaving halves hinged on one side

**Carcass weight:** The weight of the carcass after butchering is complete

**Collagen:** A fibrous protein found in connective tissue, bone, and cartilage

**Creatine phosphates:** Amino acid molecules that are an important energy store in skeletal muscles and the brain

**Cross-contamination:** The transfer of harmful bacteria from one food to another, particularly involving raw meats, vegetables, cutting boards, and utensils

**Cubed:** Tenderization using a machine with two sets of sharp pointed disks that score or cut muscle fibers without tearing

**Cure:** Any process to preserve meats or fish by salting or smoking

**Dehydrator:** An appliance designed to remove water from foods

**Dressing percentage:** The proportion of the live weight that remains in the carcass of an animal, sometimes referred to as yield. Carcass weight ÷ live weight x 100 = dressing percentage.

**Dry rub:** A spice and/or herb mixture added to the surface of foods before cooking

**Epimysium:** The sheath of connective tissue surrounding a muscle

**Extruder:** A mechanism through which food is pushed to create shapes

**Fabrication:** The deconstruction of the whole carcass into smaller, more easily used cuts

**Field dress:** To remove the internal organs of hunted game animals, or fowl

**Fillet:** To slice meat from bones or other cuts. Also, boneless slices of meat that form portion cuts.

**Flank:** A cut of a quadruped that includes the abdominal muscles

**Forequarter:** The anterior portion of a beef side, including ribs 1 to 12

**Foresaddle:** Unsplit forequarter of a veal or lamb carcass

**Freezer burn:** Discoloration of meat due to loss of moisture and oxidation in freezer-stored meats

**Fright or flight response:** A behavioral reaction by animals to a stressful or threatening situation that increases heart, lung, and muscle activity

**Gambrel:** A frame used by butchers for hanging carcasses

**Glycogen:** A polysaccharide produced and stored in animal tissue, especially in the liver and muscle, and changed into glucose as the body needs it

**Grade:** A designation that indicates quality or yield of meat based on standards set by the United States Department of Agriculture

**Grinder:** A mechanical device that crushes and breaks up meat pieces into small fibers

**Hindquarter:** The posterior portion of the beef side after the twelfth rib

**Hindsaddle:** Unsplit hindquarter of a veal or lamb carcass

**Jerky:** A nutrient-dense meat- or soy protein-based food product that has been made lightweight by drying

**Leaf fat:** Fat lining the abdominal wall in pork, commonly called kidney fat in beef and lambs

**Liquid smoke:** The substance produced from smoke that is condensed, then cooled to form a concentrated liquid for use in flavoring foods

**Live weight:** The weight of a live animal at the time of purchase or harvest

**Marbling:** Streaks and veins of fat interlacing meat cuts

**Marinate:** To soak foods such as meat or vegetables in a liquid flavor mixture (or marinade)

**Muscle pH:** The acidity or alkaline level in the muscle. It generally declines after harvest, the rate which affects meat quality.

**Mutton:** Meat from mature sheep carcasses that are usually identified by the absence of break joints

**Oleic acids:** An oily, unsaturated fatty acid present in most animal and vegetable fats and oils

**Omega-3:** A family of healthy unsaturated fatty acids

**Omega-6:** A family of unsaturated fatty acids that may increase disease and depression

**Palmitic acid:** A colorless, crystalline, saturated fatty acid found in animal fats

**Perimysium:** Connective tissue covering and holding together bundles of muscle fibers

**Petcock:** A small faucet or valve for releasing gas or air

**Posterior to:** Toward the rear of the carcass; behind

**Primal or wholesale cuts:** The large subdivisions of the carcass that are traded in volume by segments of the meat industry

**Render:** To melt down fat

**Retail cuts:** Subdivisions of wholesale cuts or carcasses that are sold to consumers in ready-to-cook or ready-to-eat forms

**Rigor mortis:** The stiffening of muscles that occurs after death as a result of the coagulation of proteins

**Round:** A muscle structure found in the rear leg of quadrupeds

**Salt pork:** Pork cured in salt, especially fatty pork from the back, side, or belly of a pig, often used as a cooking aid

**Shelf safe:** The ability of preserved or processed foods to survive long periods on home or store shelves without spoiling

**Side:** One half of a meat animal carcass

**Shrinkage:** The weight loss that may occur throughout the processing sequence due to moisture or tissue loss from both the fresh and the processed product

**Silver skin:** The thin, white, opaque layer of connective tissue found on certain cuts of meats, usually inedible

**Smoke:** To flavor, cook, or preserve food by exposing the surface to smoke from a smoldering wood material

**Subprimal cuts:** The subdivisions of wholesale or primal cuts to make handling easier and reduce the variability within a single cut

**Suet:** The raw fat found around the kidneys and loins in beef that is used to make tallow

**Yield:** The portion of the original weight that remains following any processing or handling procedure in the meat-selling sequence. It is usually quoted in percentages and may be cited as shrinkage.


# ACKNOWLEDGMENTS

I wish to thank my wife, Mary, for her constant support and encouragement. Her comments and suggestions always improve my work.

To Erik Gilg, Associate Publisher, who saw the need for this book. Thank you for the opportunity to bring this information to a larger audience.

A sincere thank you to my editor, Meredith Quinn, for shepherding the manuscript through the publication process.


# ABOUT THE AUTHOR

Philip Hasheider is a fifth-generation farmer currently raising pasture-grazed Red Angus beef with his wife, Mary, in South Central Wisconsin, from where he writes. He has had a career as a dairy farmer and an assistant cheesemaker, and, as an active historian, he has combined history with his agricultural knowledge to write twenty-seven books. His diverse writings and essays have appeared in numerous local, regional, national, and international publications and story collections. He was the writer for the 2008 *Wisconsin Local Food Marketing Guide* for the Wisconsin Department of Agriculture, Trade, and Consumer Protection that received the 2009 Wisconsin Distinguished Document Award from the Wisconsin Library Association and the national 2010 Notable Government Documents Award from the American Library Association. He is a four-time recipient of the Book of Merit Award presented by the Wisconsin Historical Society and Wisconsin State Genealogical Society. *The Ultimate Guide to Butchering, Smoking, Curing, Sausage, and Jerky Making* is his twelfth book for The Quarto Group, with other titles including *The Complete Book of Butchering, Smoking, Curing, and Sausage Making; The Hunter's Guide to Butchering, Smoking, and Curing Wild Game and Fish; The Complete Book of Jerky*; and *The Complete Book of Pork Butchering, Smoking, Curing, Sausage Making, and Cooking.* He has also penned how-to books for raising livestock, such as *How to Raise Cattle, How to Raise Pigs,* and *How to Raise Sheep.*

# GLOSSARY

**T**



**V**



**W**



**Y**

Printed in the USA
CPSIA information can be obtained
at www.ICGtesting.com
JSHW070356240924
69862JS00003B/4

9 781558 329874